Business Process Management

Business Process Management

Process is the Enterprise

Sandeep Arora

Copyright Notice

Copyright 2005 Sandeep Arora. All rights reserved. Except as permitted under the current United States Copyright Act, no part of this publication may be reproduced or distributed in any form or by any means, or stored in a database or retrieval system, with-out the prior written permission of the publisher.

Books website http://www.bpm-strategy.com
Sandeep Arora's email Sandeep@bpm-strategy.com

This book is dedicated to my wife Monika and daughter Tisha. This book is also dedicated to my father Subhash Chandra and my mother Ashok Bala.

Table of Contents

Preface 10
1. The Extended Enterprise 11
    Business Change
    The Process is the Enterprise
    Business Values
    Process Oriented Mindset throughout the Enterprise
    Process Oriented IT Infrastructure

2. Current IT and Business Divide...23
    Technology and business processes
    IT Applications and Infrastructure
    Bridge the business and IT divide

3. Business Process Management Systems..34
    Workflow Automation
    Enterprise Application Integration
    Collaboration
    Working without a process mindset using BPM tools
    Business Process Management Systems (BPMS)
        Business Process Life Cycle
        BPM Infrastructure
        Work or Task Portal
        User Directories
        Business Rules Engine
        Enterprise Application Integration Capabilities
    Benefits of BPM

4. Business Process Design..46
    BPM Team
    Process discovery and design
    BPM Terminologies
        BPD (Business Process Diagram)
        Process Instance
        Process Participants
        Process Activities
        Work Items (and Work Lists)
        Sub-Processes (Or Nested Process)
    BPM Tool Requirement – The Process Designer
        Intuitive process flow design
        Multiple process repository support
        EAI Activities

    Multiple Form Support
    User directories and routing
    Nested Model Support (Sub Processes)
    Collaborative Design
    Process Simulation
    Event Notification Services
    Process documentation
  Business process design best practices
    Event Notifications (Alerts and Notifications)
    Work in parallel
    Web services
    Sub Processes
    Exception Handling
    Cluttered business process design
    Interdependent Activities

5. Business Process Deployment 67
  BPM environments
  Process testing in QA
    New Process Deployment
      Process Flow Paths
      Sub Processes:
      Alerts and Notification:
      Automated Activities:
      EAI (Enterprise Application Integration) activities
      Exception Handling Testing
      User Directory Testing
    Updated Process deployment
      Deployment Strategy
      Process Downtime
      Deploy Updated Process (continuous business operations)
      Side-by-Side version testing
  BPM Pilot
  Process Deployment to Production

6. Business Process Monitoring and Exception Management 80
  Business Process Monitoring
    Alerts and notifications
    Business Intelligence
  Exception Management
    Business process exceptions
    Participant Related Exceptions
    Anticipate/Avoid Exceptions

Detect Exceptions
Exception resolution process

7. Process Optimization 93
Business dynamics and change
Inputs from Exception Management and Process Monitoring
BPM and the Balanced Scorecard
    Process metric improvement
    Process efficiency metrics
    Process visibility metrics
    Cost effectiveness metrics
    Resource planning metrics
Process Optimization

8. EAI (Enterprise Application Integration) and BPM......................106
EAI (Enterprises Application Integration)
EAI in BPM
Web Services
BPM and Web Services – A powerful combination

9. Collaboration..114
BPM and documents
Discussion Threads
    Discussion groups around the live process instance
    Discussion groups around activity instance
Real-Time Collaboration
BPM specific collaboration
    Returning work items
    Sending work item for reference

10. BPM Servers..122
Orchestration Services
State Management
Transaction Services
Management services
Business rules services
Enterprise Application Integration services

11. Process Excellence..130
Flextronics
UPS (United Parcel Service)

Appendix A: BPM product greatest needs. ……………………..136

References………………………………………………………….141
Index ………………………………………………………………..142
About the Author ………………………………………………….144

Preface

In today's information and knowledge based economy, process is the enterprise. Companies serve customers by executing business processes. Companies have innovative ideas and strategies which they want to execute quickly, before opportunity is lost. They have quality initiatives like six sigma, business process automation, balanced scorecard, and activity based costing which they would like to implement. Before BPM the technology did not exist which enabled companies to run digitized business processes and measure performance in real time.

But there has been a lot of hype and confusion regarding BPM. I have been an early adopter of BPM and have implemented enterprise strength mission critical BPMS. Based on my experience becoming a process managed enterprise is challenging and requires hard work. It requires a mindset change for all (business and IT).But the rewards of being process managed are tremendous and the effort is worth it. Process Managed enterprises change the basis of competition.

In this book I unfold my lessons and experiences regarding BPM from a higher level. This is the first book by someone who has been working with BPM at the grass roots level, so I do not paint a false picture. Based on real experience, I also explain why BPM offers revolutionary advances for process management. I believe companies which adopt BPM will rule their respective industries in the years to come.

The following are the goals of the book
- Importance of BPM and why BPM cannot be ignored.
- Evolution of BPMS from existing technologies like EAI, Workflow and Collaboration.
- Setting the correct process mindset for enterprise stakeholders, BPM implementers and users.
- Best practices and strategies for all phases of process life cycle (based on real project experience).
- BPM product greatest needs and evaluation.

My goal is to be a proponent of BPM.

Sandeep Arora
April 2005

Chapter 1:

The Extended Enterprise

Extreme business-cycle volatility is norm in today's economy. The pace of business change has been accelerating over the past few years. Businesses have to deal with an array of business change forces. There is no way to insure your business against change. The way to insure and hedge your business against change is to change constantly. Businesses must change in order to prosper. Changes in government policy, global economy, rules & regulations, corporate mergers and acquisitions, demographics, competition and technology are some of the forces that force business to be Agile.

Business Change

Here are some real time examples of Business Change happening today. Below factors will make it clear that the rate of business change is accelerating and it is happening now.

Government Policy and Rules & Regulation
The Federal Sarbanes-Oxley Act of 2002 known as SOX establishes rules effecting corporate governance, financial disclosure, and the practice of public accounting. Companies have to comply with SOX thus making changes to existing business processes and gaining visibility into their business practices. Most enterprises do not have complete visibility into their business processes. As a result they are struggling to comply with Sarbanes-Oxley Act and spending huge amount of resources on

- Documenting and updating changes to the Company's existing business, accounting and financial reporting processes
- Identify existing control points in the Company's processes, as well as "control gaps" requiring remediation

Clear visibility into corporate finances, regulatory fever after all the financial scandals we have seen and coupled with ongoing efforts to make businesses more productive with better information is the call of the enterprise today. Business Managers need to see how companies are performing in real time and make strategic decisions.
The New York Attorney General Eliot Spitzer launched the first salvo against alleged conflicts of interest, charging the insurance brokerage arm of Marsh & McLennan (MMC) with price fixing and collusion. Some of the nation's largest insurance companies where accused in Spitzer's suit of steering contracts and bid rigging. As a result of this investigation, the New York office has requested mountains of information for the past decade or so from many insurance companies. CEO of Mass Mutual, said on a TV interview that they had been contacted for all information they had, dating back several years. To accomplish this Mass Mutual

had several resources gathering and preparing the information needed. He mentioned that for a smaller company it would have been a drain on resources.

Companies which have control, visibility of their business, financial and accounting processes are better positioned to handle and adapt to an ever changing Landscape of Rules and Regulations. Companies which have poor control and visibility get distracted, spend valuable time and resources trying to comply and loose market share to competition.

Customers

There has been a shift in the balance of power away from the companies and towards the customer. Thanks to the internet customers have become more demanding to a point of being ruthless. Changing customer demands and habits are reshaping entire industries. Case in example is the music Industry. The delivery, consumption, cost structure, storage and distribution of music has changed and has reshaped the entire music industry. In a few short months Apple has become the leader with the cool **iPod** product and **iTunes**. Customization i.e. tailoring products and services to the specific preferences and needs of customers without sacrificing cost, efficiency and flexibility is being expected out of all companies. We are in an age of mass customization.

The "self service" customer wants to book his own flight, rent a car online, manage his finances in the middle of the night, cancel his hotel reservation, and buy a pair of jeans from the comfort of his home. Basically the customer wants to sit in your boardroom and dictate business strategy. Business change could be your biggest problem or your greatest opportunity. In a global economy and shorter business cycles companies need to connect to their customers, suppliers and partners in real-time. That way they can respond quickly to changes.

> **Eric von Hippel, professor at MIT's Sloan School of Management, writes about customer centered innovation in a new book, "Democratizing Innovation", which states that manufacturers and producers should redesign their product design processes to seek out and include user ideas and innovations.**

Von Hippel says that by drawing and capitalizing on the creativity of "lead users", who are often ahead on technology and domain expertise, companies can improve the chances that their products and services are commercially viable. Basically these trends, like the "open source" development model of the software field clearly indicate the customer or the end user is a dictator.

Technology

Rapid advances in technology is creating new industries and making some industries obsolete. Case in example is RIFD (Radio Frequency Identification) technology. RIFD is a generic term used to describe a system that transmits its identity wirelessly using radio waves. You can get more information by visiting http://www.rfidjournal.com . Wal-mart wants its top 100 suppliers to use RFID tags on cases shipped to some of its distribution centers by June 2005. This action has sent suppliers and competitors scrambling to learn about the wireless technology, which enables companies to identify and track items in the supply chain automatically.

"Bar codes have transformed the way we all do business" said Mike Duke, president and CEO of Wal-Mart Stores Division (USA). "RFID will not just transform how we do business but will revolutionize how we all do business. I don't think we can even imagine all the benefits that it will deliver." RFID technology is certainly going to change the way companies do business – and companies which are slow to adapt will have to face consequences. The decline in prices of tags is going to trigger a mass adoption – probably making the bar code obsolete in the next 2 decades. Such tsunami business changes do require business to be agile.

It took Coca-Cola several years to create a global brand world wide in the industrial age. But in the digital age it took Google couple of years to become a global brand. Non-existent 6 years ago, it's now part of our language, as in "Google me", or "Google that". To Google – means to get an instant answer by using the company's search engine to search anything on the internet.

Globalization

Globalization is happening now at a break neck speed. China has become the world's production factory. That has forced many businesses to completely change their business models. Companies which could not adapt and get manufacturing done in China are out of business. If you go to Wal-mart store today and pick up any product – chances are it is made in China. US has been losing manufacturing industries to China for decades now. Certain industries like apparel are wiped out from the USA.

Advancement in telecommunication over the past couple of years has enabled businesses to tap talent anywhere in the world. Leveraging global knowledge asset is the new mantra. Business process outsourcing is becoming the strategic choice of companies looking to increase efficiencies, reduce cost, expand into new markets and focus on there core business. Advances in telecommunication technologies is forcing the creation of new industries and making some industries entirely obsolete. Call Center's, R&D centers and development centers are being set up in India by corporations. India is quickly turning out be the backoffice to the world.

In today's business environment one company cannot produce everything needed for the end product. Suppliers, partners, customers are involved in the value chain. We are talking about the "**Extended Enterprise**". The driver or the central nervous system of the extended enterprise is the business process.

Competition

Your customers are a click away from your competition. Competition in all industries is cut throat, and with the internet customers have all the information at their finger tips. Comparison shopping using the internet is norm. Most customers do channel hopping before making a decision to buy a product or service. Many of us research a product on the internet, see the physical product in a brick and mortar store and eventually buy it online at Amazon. Customer loyalty is pretty much dead.

Peter Fingar author of "The Real-Time Enterprise" says companies need to "Compete on Time". He writes "*Cutting-edge companies today are capitalizing on time as a critical source of competitive advantage: shorting the planning loop in the product development cycle and trimming process time in the factory – managing time the way most companies manage costs, quality or inventory. In fact, as a strategic weapon, time is the equivalent of money, productivity, quality, even innovation. While time is a basic performance variable, management seldom monitors its consumption explicitly-almost never with the same precision accorded to sales and costs. Yet time is a more critical competitive yardstick than traditional financial measures.*"

Process Efficiency

In today's cutthroat environment and competition companies need to be extremely efficient. The continuous ability to improve productivity, efficiency and cut costs needs to be at the core of an enterprise. The prices of computers have been declining for the past many years but Dell continues to profit in the face of shrinking margins. Dell is a perfect example of a company which continues to increase the efficiency of its direct model process. At the core of Dell is the super efficient, extremely agile process. IBM could not imitate Dell's process model and had to eventually get rid of their PC business.

The Process is the Enterprise

> The tsunami wave of business change is disrupting the business landscape, reshaping certain industries/companies, destroying certain industries/companies and creating new industries/companies.

Business conditions are changing at a rapid pace. Working in the same business conditions –some companies continue to thrive while others get caught in the tsunami wave of business change and struggle for survival. What differentiates them is the way these companies get the job done- that is the called the **business process**. A business process is comprised of the people who work, the tools they use, the procedures they follow, the methodology they use and the flows of material and information between the various people, groups, systems and sub-activities.

> **A business process is a complex function of several variables like culture, organization structures, organizational practices, method of work, structure of work, employee skills, flow of information and technologies.**

A study was done by MIT Sloan School of Management in a paper titled "*Intangible Assets: Computers and Organizational Capital*" by Erik Brynjolfsson, Lorin M. Hitt, and Shinkyu Yang to establish a link between these intangible assets (which I collectively refer to as business process), computer assets and market value. Below is the three dimensional graph (published in the above paper) which came out of the survey done by contacting several hundred large firms. The research indicates that in a digital age spending on tangible assets like IT alone is not sufficient to be a market leader. IT spending has to be coupled with spending resources on intangible assets like process orientation, business processes, organization structuring, etc to achieve process excellence.

The paper (*Intangible Assets: Computers and Organizational Capital*) mentions that "*the truly valuable assets were the complementary business processes, work practices, and even culture, all of which were harder to identify and implement. In effect, these constituted an organizational asset with real value, although one not reflected on the firm's balance sheet.*" The author says and as shown Figure 1.1(from the above paper) is that they found that the intangible assets (organization structure, business processes etc) complemented by IT assets increase productivity many times over than just investment in IT.

The report mentions that "*Firms choose to invest in certain business models, organizational practices, and corporate culture. Later some of these investments turn out to be more productive and profitable than others. The financial markets recognize and reward those models that are well suited for the current technological and business environment. At that point, other firms may try to imitate the winners' best practices, but the complexity due to explicit and implicit complementarities among each collection of practices makes this difficult. Kmart may wish it could emulate Wal-Mart, and Compaq may try to learn from Dell, but their adjustment costs may prevent this from happening for years, even if they succeed in the end.*" It is pretty clear that business processes are the differential factor in today's dynamic business environment.

Chapter 1: The Extended Enterprise

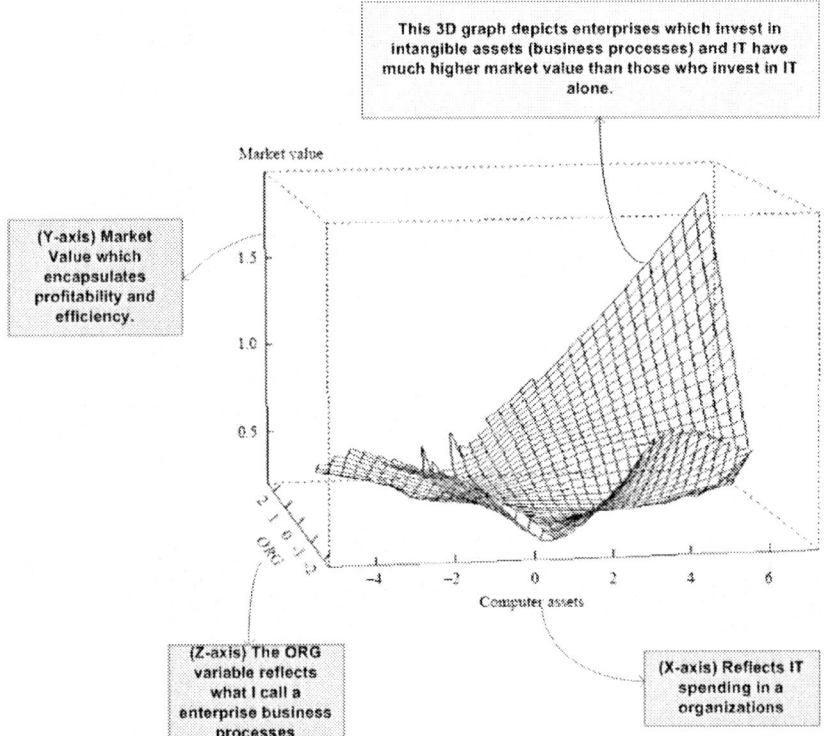

Figure 1.1 Relationship between computer assets, business processes and market value.

Business process is king .The process is the enterprise. Business process can extend across the enterprise connecting customers, suppliers and partners.

- The X-axis in Figure 1.1 represents the IT spending done by the surveyed companies.
- The Z-axis which goes inside the graph is represented by the ORG variable. This is an intangible organization asset. The following components were used to construct this variable
    - Degree of team self-management
    - Employee involvement groups
    - Who determines pace of work
    - Who determines method of work
    - Degree of team building
    - Workers promoted for teamwork

- ➤ Percent of workers that receive off-the-job training
- ➤ Degree of screening prospective hires for education
- The Y-axis is the market value indicator.

IBM, Dell, Gateway, HP and Compaq are all in the PC business. Dell continues to prosper and the rest are struggling in their personal computer businesses. The only differentiator between these companies is the direct model business process of Dell. IBM recently sold off its personal computer business to a firm based in China named Lenovo. HP acquired Compaq and Gateway is struggling with loses. It is pretty clear that the business process is the differentiator here. All these firms could not copy Dell's business process. I would safely and with confidence say that in today's information age and knowledge based economy **"Process is the Enterprise"**. Dell has continuously improved its direct model business process from a cost, efficiency and quality standpoint. Dell is an example of a great **"process managed enterprise"**. Excellence in process management enables companies to execute business strategies.

Wal-mart is the perfect example of **"process is the enterprise"**. Retail industry is the most competitive industry in the world, but Wal-mart with annual sales > $250 billion dollars continues to grow. Wal-mart does not manufacture any of its products. Wal-mart uses it's highly flexible, efficient and agile supply chain to orchestrate its suppliers and partners around real time information. Wal-mart is on a relentless drive to improve efficiencies and drive cost low.

- Anytime a customer buys a product at any Wal-mart store – the point of sales transaction data reaches the central computer data center in one hour.
- The customer data is analyzed for customer and demand analysis.
- Wal-mart has opened up its books to external suppliers like P&G, DreamWorks, etc so that they can monitor the sale of their products in all Wal-mart stores in real-time. That way suppliers can make real-time logistical decisions to supply products to Wal-mart.
- Weather tracking is part of Wal-mart's supply chain process. Wal-mart's supply chain moves based on weather events. A winter snowstorm forecast in Boston area will result in more snow shovels, salt, snow blowers, water, snow shoes, and snow related products available at Boston area stores.

It is evident that the acceleration in business change has changed the field of competition. The competition in an extended enterprise is between business processes. What matters now is the ability to control, monitor, enhance and improve with great efficiencies your end-to-end business process. It does not matter whether or not the processes and sub-processes reside within the company's organization. Business need to make their process more elastic and their organizations more flexible. In the extended enterprise functions and departments are outsourced to partners and suppliers with a clear set of instructions.

Business Values

Companies need to manage their business processes in the ever changing landscape of the business. Companies need the process mindset, organization structure, process culture, tools and technologies to manage their business processes in the face of market changes happening now. Table 1.1 is a list of attributes needed by an enterprise today. All the previous initiatives like TQM (Total Quality Management), BPR (Business Process Re-engineering) and Six Sigma have one message in common "***Business processes are key and managing the business process is vital to the health of the enterprise***".

The recent coverage given to business processes in management and technical magazines, conferences and reports has created a good awareness of business processes and business process management. It has also created hype and confusion regarding which technologies and methodologies enable process management. Many senior executives are now aware of the need to manage enterprise business processes – but they do not know how to implement it. To be a process managed company is about managing the value chain with agility, visibility, flexibility and efficiency.

> **Process managed enterprises do not have synergy gaps between strategy conceptualization and strategy execution.**

Michael Hammer, co-author of "Reengineering the Corporation", is his book "The Agenda" writes "*No matter how hard individuals work, they cannot overcome a flawed process design, much less the burden of no design at all.*" Enterprises serve their customers needs by providing value to customers by knowingly or unknowingly executing a business process such as "Sell product to customer", "Customer support", "Develop market" etc. without regard to the individual functional departments that might be involved. To be a process managed company means having a process oriented mindset and having the methodologies and tools to manage the business process. Working with a process oriented mindset is about organization culture change. I will not be going into much detail regarding people and culture issues as that is not the objective of my book. Andrew Spanyi in *Business Process Management (BPM) is a Team Sport: Play it to Win!* proposes that organizations need to consciously work on transforming the mental models of the executive team from the traditional functional paradigm to a customer-driven model that is based on business process thinking.

Agility	The ability of a business process to continuously adapt to ever changing demands of the market place. Agile companies can add/modify products and services quickly, and can respond to the other forces of the market without going out of business.
Visibility	The ability to see where the business is in real-time. Visibility enables business managers to make better decision, comply with rules and regulation, measure key performance indicators, enhance customer service, etc.
Flexibility	The enterprise should be able to deal with business exceptions quickly and efficiently. The ability to transform in the face of change, leverage existing infrastructure and resources.
Efficiency	The ability to continuously improve the speed and drive down the cost of a Process without cutting the quality of the product or service.

Table 1.1: Desired Business Values in a ever changing business landscape

Most enterprises are structured and managed by divisions and departments that are dedicated to performing specific functions and have personnel who are experts at those functions. Employees view themselves based on their position in the organization chart. But a customer's view is that of a service provided by the enterprise. A customer does not view an enterprise as a collection of departments. A customer views the enterprise through the business process by which he is served.

Local process improvements rarely translate into enterprise wide productivity or efficiency gains.

Figure 1.2 depicts a functional view and a process view. Dividing tasks vertically and optimizing departmental way of working is not the best way to optimize the end-to-end business process. The business process runs horizontally, it crosses and re-crosses organizational departments, suppliers and partners. The functional approach creates bottlenecks for the process as there is the risk for task duplication, loss of quality and delay. Some organizations do have pockets of effective business process best practices based on management theories like TQM (Total Quality Management), BPR (Business Process Re-engineering) but they are not linked and do not sum up to being a process managed enterprise. BPM (Business Process Management) is a methodology to manage the enterprise wide business processes. BPM is a way of working and an incremental approach. I will be covering BPM in detail in next chapters.

Chapter 1: The Extended Enterprise   21

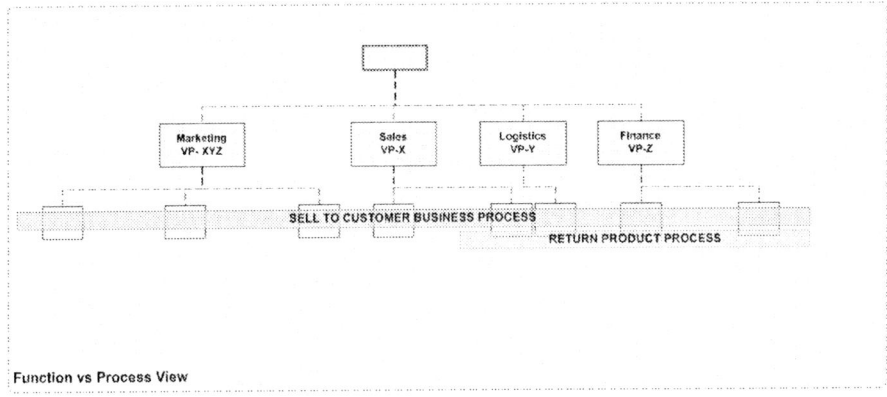

Figure 1.2 – Processes run horizontally crossing departments.

Process orientation throughout the Enterprise

When a customer applies for a mortgage – he views the company as a mortgage provider, not as a collection of departments – sales, underwriting, appraisal, etc. The customer does not want to be thrown around all these departments and he does not want to interface with all these departments. The customer wants to give his information once and not multiple times. Customer wants the mortgage approval process to be smooth, transparent and the customer wants to be in control. But the reality is different as all of us know. Refinancing or getting a new mortgage can be a nightmare.

Business people need to start managing their enterprise from a process standpoint. They need to look at their business processes outside in. That means a new way of working. To implement BPM the senior management at the highest level needs to understand why they are doing it, the pain associated with change and rewards of being a process managed company. Implementing BPM needs to be a corporate objective. Working with a process mindset is painful for business people who are conditioned to working in a functional mindset due to many reasons:

- Unlearning traditional ways of working.
- Learning new ways of working.
- Loosing some or all control of information.
- Increasing visibility which is tied to people feeling less important.
- Being accountable which means people activities are monitored using BPM systems.
- Automation and digitization sometimes means reduced workforce.

Organization change is the major hurdle in successful BPM implementations. But on the other hand employees can enhance their business expertise with a process oriented mindset.

Process Oriented IT Infrastructure

Once an enterprise in mentally prepared to become process managed the next step is having the right methodology, mindset, framework, tools and IT infrastructure. Over the past three decades organizations have created several enterprise systems. At the time these applications where created the purpose was served. Computers and software applications have improved productivity. Databases have helped organization manage large volumes of business data. But with the Internet which is a universal connector – anywhere, anytime – the landscape has changed and coupled with globalization, information everywhere, competition, falling prices etc the pace of change has accelerated.

The IT applications and infrastructure has not kept up with the pace of business change. In today's ever changing business environment these systems have been reduced to mere data collection and data storage systems. Most of them are function specific systems. ERP, CRM, Peoplesoft, etc are all function specific applications with some amount of automation built into them. As a result we see the business and IT divide which I will write about in the next chapter.

What we need today is process oriented IT infrastructure. We need the tools and methodologies for process management. Up until now IT applications have been involved in parts of the business process. What we need now is technology to manage the complete end-to-end business process.

Chapter 2:
Current IT and Business Divide

Technology and business processes

Technology is just an enabler of productivity, efficiency and cost reduction. Businesses cannot get productivity gains and become process managed just by getting the latest and greatest technology. Business need to work hard and change to become process managed. Technology needs to be weaved into the organization business processes to achieve maximum efficiencies and agility. As we saw in Figure 1.2 Information technology is just one part of the axis when it comes to being a process managed enterprise. The real killer apps of the past decades have been super efficient and super agile business processes like the supply chain process of Wal-mart and direct model process of Dell – technology has just been a catalyst for this kind of process innovation.

IT Applications and Infrastructure

Most of today's IT applications and enterprise systems have business process and business rules built into them (Fig 2.1). Business processes and business rules are everywhere including people's heads. The rate at which business process changes is so rapid that making corresponding changes to IT Applications that support them is virtually impossible. As a result the IT applications which once supported the business fairly well at the time they where built are out of sync. Software development life cycle is way too slow compared to business process life cycle

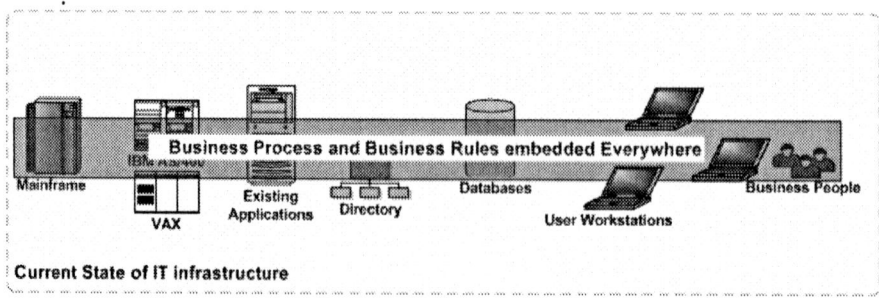

Figure 2.1 Business processes and rules are everywhere.

Business data has been collected over the years –thanks to RDMS (Relational Database Management Systems). Databases are excellent in capturing and retrieving data. Databases have allowed companies to capture complex and vast amounts of data in a relational fashion. RDMS have matured over the years and are very efficient and reliable – easily supporting millions of transaction per day. As a result vast amounts of data have been created over the years. Data warehousing tools are used to analyze this data to research trends and forecast demand. In most organizations managing this vast amount of data and deriving any business value out of this huge amount of data is the greatest problem.

Most organizations are department based as we saw in chapter 1 and have different view of the same data. The sales department and marketing department of the same organization might have different view of the customer data. As a result data quality suffers. In one of my consulting assignments with a global financial institution different departments had different definition and storage formats for "*perspective clients*". The pace of business change is making this data irrelevant sooner than it had in the past. Forecasting done against obsolete data can result in bad business decisions. Putting this data in a real time process context is extremely challenging as this data resides in multiple repositories.

Since the early 1990 there has been a lot of hype about EAI (Enterprise Application Integration) and B2B (Business-to-business) integration as a solution to all business process and data problems. EAI platforms replace a lot of "point-to-point" integration. EAI projects focus on data and are extremely complex and expensive to implement. Later packaged applications like CRM (Customer Relationship Management), SCM (supply chain management) and ERP (enterprise resource planning) have emerged as good standalone department specific systems. Howard Smith co-chair of Business Process Management Initiative (BPMI.org) says "*These packages implemented best-practice processes but did so by ingraining business processes in the software applications that supported them. These solutions had all the flexibility of wet concrete before they were installed and all the flexibility of dry concrete after installation.*"

Enterprises have created pockets of automation and numerous data repositories. All these applications, automation and data repositories do not deliver an efficient and agile end-to-end business process management solution. Current IT infrastructure does not address the fact that business process will continue to change at a rapid pace. Most organizations today have what I call a rigid and brittle IT infrastructure. Organizations which have rigid and brittle IT Infrastructure

- Difficult to improve business processes.
- Cannot easily convert strategy to execution.
- Cannot respond to market changes.
- Are not efficient.

Rigid IT infrastructure renders all the business processes it serves rigid as well. Figure 2.2 depicts the problems and the way work is done in an enterprise with a rigid and brittle IT infrastructure. Any change to one system effect's an array of systems and people. Figure 2.2 also shows the way most organizations work with people from each department having to go to multiple systems to enter/retrieve information. Rigid IT makes it extremely difficult for companies to deal with business change.

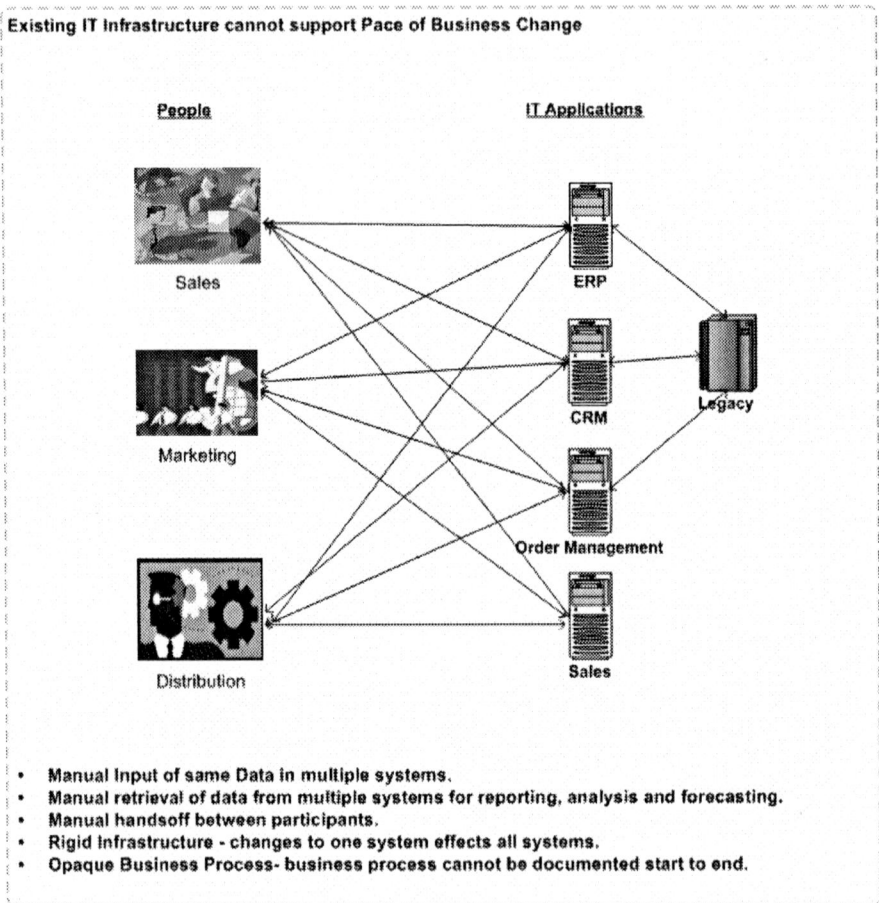

Figure 2.2 Problems associated with rigid and brittle IT infrastructure

> **Rigid business process :** A business process which is extremely expensive and resistant to change by virtue of pockets of automation, point to point integration and business rules and structures spread across various IT application layers.

Nicholas G. Carr in his highly controversial paper in Harvard Business Review (May 2003) title "IT Doesn't Matter" writes regarding the thinking that organizations can increase productivity and adaptability by making investments in IT *"that as It's potency and ubiquity have increased, so too has its strategic value.*

It's a reasonable assumption, even an intuitive one. But it's mistake. What makes a resource truly strategic – what gives it the capacity to be the basis for a sustained competitive advantage – is not ubiquity but scarcity. You only gain an edge over rivals by having or doing something that they can't have or do. By now, the core functions of IT – data storage, data processing, and data transport – have become available and affordable to all. Their very power and presence have begun to transform them from potentially strategic resources into commodity factors of production. They are becoming costs of doing business that must be paid by all but provide distinction to none."

I think he meant investment in hardware and software is not enough. Investment (time and resources) need to be made in business processes which ultimately matter. Knitting IT into an innovative and efficient business process is what makes an organization unique and a leader in its field.

> **Business processes are fragmented across functional areas and applications thereby creating errors and inefficiencies. Sub-optimal process communication methods include memos, faxes, email and phone, which leads to many manual, error prone steps, redundant steps and wasted time and effort. There is no visibility in this process. Paper trail cannot be followed. This has resulted in lack of consistency, bottlenecks, rigidity, poor quality and workarounds.**

Business managers turn to IT to provide them with information and tools to manage the business. But the array of enterprise applications does not deliver end-to-end process management. The business managers want IT to be a strategy execution machine. But the reality is IT has become a data collecting cost center. As a result there is very little alignment between the business community and IT. Businesses need to rapidly respond to changed business conditions or capitalize on new opportunities. Businesses which fail to respond could find themselves on the brink of extinction. The inability to rapidly translate new business requirements to IT applications is a major roadblock to the company.

Business Community is always saying "Why will this small change take so much time? All I want is a report with data coming from two systems to manage my clients". I'm sure most of us have said or heard this statement many times. Business managers cannot wait months for the business requirement changes to be implemented in the array of IT applications. So out of frustration, when there is a change in the business process the business users either ask IT to piecemeal it or they start creating excel spreadsheets and word documents with business process rules and data for immediate use. Overtime these spreadsheets and documents become mission critical and part of the business process. Business process rules and business specific information is also mailed around for immediate action. So basically business intelligence is buried in spreadsheets, word files and emails everywhere.

Value Add Task
The task that directly contributes towards what your customer wants out of the product or service and the customer is willing to pay for it.
Non-Value-Add Task
Task that adds no value in the eyes of the customer and customer does not want to pay for it like • Task requiring multiple manual approvals. • Multiple input of information into various systems. (Fig 2.2) • Tasks requiring retrieving and analyzing data from multiple systems (Fig 2.2) • Redundant activities
Non-Value-Add (Business) Task
Tasks that customers do not care about and do not want to pay for, but are required from a business perspective like • Task required for federal, state or global compliance. • Accounting task. • Task required to reduce financial risk.

The end result is that many **Non-Value-Add** (see above) task are needed to complete the business process. Some **Non-Value-Add (business task)** are required by law for the business to operate. The business process becomes extremely slow, error prone, **opaque**, difficult to track, cost ineffective and even more resistant to change. Figure 2.3 represents the business process as it is done today. There are too many non-value-add tasks which drain company resources. There is absolutely no way to track a work in progress, very difficult to improve and very cost ineffective.

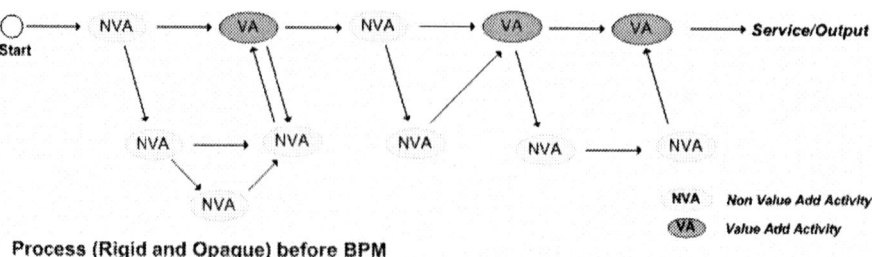

Process (Rigid and Opaque) before BPM

Figure 2.3 Business process execution as it is today with process management.

> **Opaque Business Process: A business process where there is a lack of visibility and lack of rules and common understanding among the process participants and process managers.**

I was consulting for a large financial institution back in 1997. When they planned a company wide Windows NT rollout the migration team counted over 400 desktop applications (Excel, Access and thin client applications) sitting on business user desktops. These small applications where created by the technology savvy business users. These spreadsheets and applications where critical to some business process or functionality area of the business. Basically in most companies business process just evolve without any planning. Till that time IT was totally unaware of these desktop applications and so these applications were not supported. Lack of common understanding of business process creeps in within the respective departments and organization as a whole.

What I have found out is overtime as people leave and join the organization, business process changes over the years, mergers and acquisitions, etc the state in which most organizations are is that they do not have a complete understanding of their business process. Same departments in different locations execute the same process in a different way. The biggest challenge many corporations face is putting on paper the "as-is" business process during process design.

Bridge the business and IT divide

IT applications on which business processes depend are rigid. Implementing change is complex, expensive and time consuming. Organizations need a holistic approach to manage their business processes. Organizations do not need another set of tools or packaged software applications. What organizations need is a set of infrastructure, technologies and methodologies that can help the business manage their business processes. Business managers need more control in terms visibility and manageability over their business processes. These technologies and methodologies should work around the process and not around data or technology. Business managers need the tools and technologies to manage end-to-end business processes in real-time.

Necessity is the mother of invention. Business Process Management (BPM) attempts to shrink the business and IT divide. BPMS (Business process management systems) are built using BPM tools. BPM enables companies to design, model, deploy and manage business processes for the extended enterprise. Born from the need to continuously manage business change BPM systems will enable corporations to manage their business processes in real time. BPMS will create actionable business intelligence in real-time not after the fact reports and analysis.

So far IT applications have been dealing with the consequences of business process change. BPM will now deal with business process change itself. I will be covering BPM and BPMS in detail in the next couple of chapter. Before that I want to lay a strong foundation on the core underlying concepts of BPM. The following concepts are the backbone of BPM:

Abstract the "business process" out of the IT applications into a separate layer called the process Layer:

- Processes will be abstracted to a process layer. The process will be expressed in a notation which is understood by both business and IT. That way the business can participate in process management.
- When business process change the process layer will need to change. The array of enterprise applications will not have to be changed as it is done today.
- IT applications will serve the business process and be process centric.

Separate the "business rules" from the business processes:

- Business rules will be managed by the business users in real time.
- A Business rules engine (BRE) is used to define and maintain the business rules.

Enterprise Application Integration should be done around the business process and not around data or IT Applications:

- EAI should be process centric – not data centric as it is now.
- This includes B2B (business to business) collaboration and communication.

Collaboration (human to human) should be done around business processes and not around IT Systems:

- Human participants need to collaborate to achieve a common goal. Collaboration should be around the business process in secure and reliable manner using a variety of synchronous and asynchronous methods.

Figure 2.4 depicts from a higher level how people will work in a process oriented enterprise. The business process is at the center with everything else revolving around it. Figure 2.5 represents the business process after process design and BPM implementation. Notice the process improvement and reduction in the non-value added task.

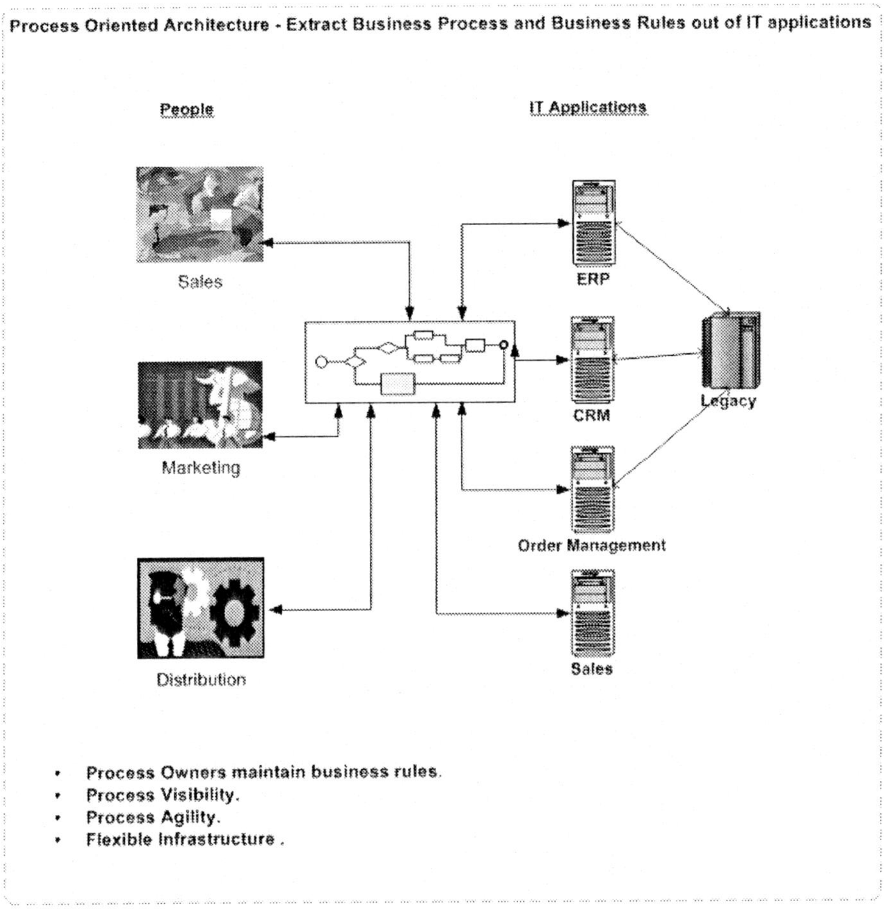

Chapter 2: Current IT and Business Divide 31

Figure 2.4 Process Orientation

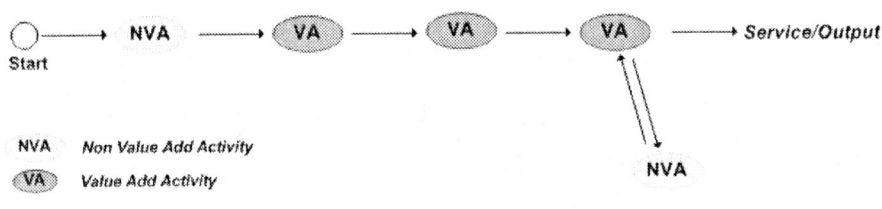

Process After BPM

Figure 2.5 Business process after business process management

> **BPM manages business change before the fact. Unlike today's IT applications which mitigate business change after the fact.**

Howard Smith author of the "Third wave- Business Process Management" writes "BPM feels similar to Computer-Aided Software Engineering (CASE) because of the emphasis on graphical notation, collaborative discovery and design". As you will read in the following chapters the business process owners are involved in every stage of business process life cycle.

> **The design goal of BPM (Business Process Management) is to manage change. BPM is about pulling the levers of cost, efficiency, quality, agility and visibility in the most optimized way.**

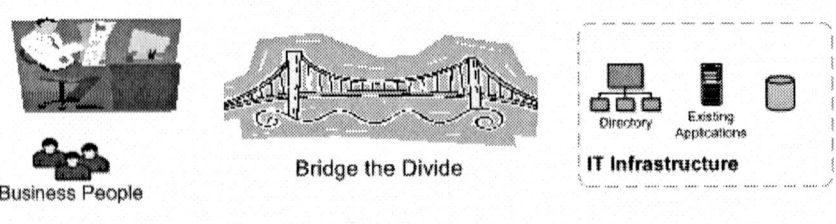

Figure 2.6 BPM bridges the business and IT divide.

IT department will be at ease with BPM. But IT will have to learn what it means to be a process managed enterprise. The IT department people will need to have a strong understanding of BPM concepts and how it is linked to enterprise strategy. The IT applications which IT develops will be participants in the business process. The IT department needs to unlearn some of the old ways of developing IT applications and learn new BPM development concepts.

This is important because IT has been conditioned with working in a data centric mindset. Working with a process centric mind set will help the IT department not make the same mistakes of creating rigid IT infrastructure. BPM projects cannot be managed as IT projects. We will see that as we move along in the next few chapters. To shrink the IT-business divide, below are some of the changes for IT

- IT department cannot work solo as in the past. IT department needs to establish strong relationship with the business units. As you will read in the next chapters business people will be involved in all phases of business process life cycle. As the relationship deepens both the IT department and business units will work as one logical entity.
- IT department focus shifts from technology to process.
- IT department is used to longer software development life cycles. They will now have to learn and embrace the shorter business process life cycle.
- Business processes are long running transactions (days, weeks or more). IT people have dealt with short running transactions (seconds or minutes). This change represents a paradigm shift from a IT development perspective.
- IT needs to be BPM savvy.

Couple of years ago I was working on a electronic business exchange project where buyers and suppliers met. It was a pure process oriented project. We (the IT team) had a data and application mindset and had no idea about BPM. We created a rigid electronic business exchange with process and business logic buried deep in the code. It was years later after learning about BPM I realized our mistake. Coming from an IT background myself I must say it is extremely important for IT department to be on the same page as the entire organization regarding BPM.

Much of the infrastructure on which BPM systems will be built is already in place. BPM products are available now and are maturing. BPM methodologies and best practices are evolving as the BPM awareness increases. But beware there is a lot of confusion out there on what exactly is a BPM system. EAI vendors are re-branding their products as BPM Tools. Workflow Automation vendors are claiming to be BPM solutions providers. Business community has understood that being a process managed company is the way to go. They have understood that to survive and thrive in an ever changing environment of business change they need to be process managed.

Business Process Management Systems (BPMS) will enable companies to manage the business process. But the question is "How the heck do we implement BPMS?" In the next chapters I plan to explain BPMS components and capabilities. I will also write about BPM methodologies to follow. I have been one of the early adopters of BPM and have built BPM systems. And believe me we built on top of what we had. The next chapter deals with the philosophical and practical aspects of designing and implementing a BPM system. Most of the BPM vendors out there speak about how easy it is to design and implement a BPM solution without programming. But the reality is anywhere close. As you will see in the following chapters, BPM implementation requires a process mindset and proper planning.

Chapter 3:
Business Process Management Systems

Chapter 3: Business Process Management Systems

"BPM (Business process management) is not just a technology. It is not a product. It is not a killer App." I would say it a "Killer way of working born out of the need to manage the continuously changing business process". BPM is a business discipline to manage the business process lifecycle. BPM is not a new management theory, but it is a new way of working. BPM includes, but is not limited to supporting Six Sigma, Balanced Scorecard, Total Quality Management, Business process re-engineering, etc. BPM products provide the set of tools and services needed to manage the business process no matter what combination of management theory you use. When a business process is automated and managed using BPM tools the end result is a BPMS (Business Process Management System).

BPMS (Business Process Management Systems) will enable the extended enterprise to manage its business process value chain across enterprise applications, customers, suppliers and partners. The BPM product will provide all the necessary services needed for end-to-end process management. BPMS is the business process management platform that orchestrates the business process with all the human and system participants giving complete visibility and control to the business managers. Figure 3.1 how BPM evolved from a management and technology standpoint.

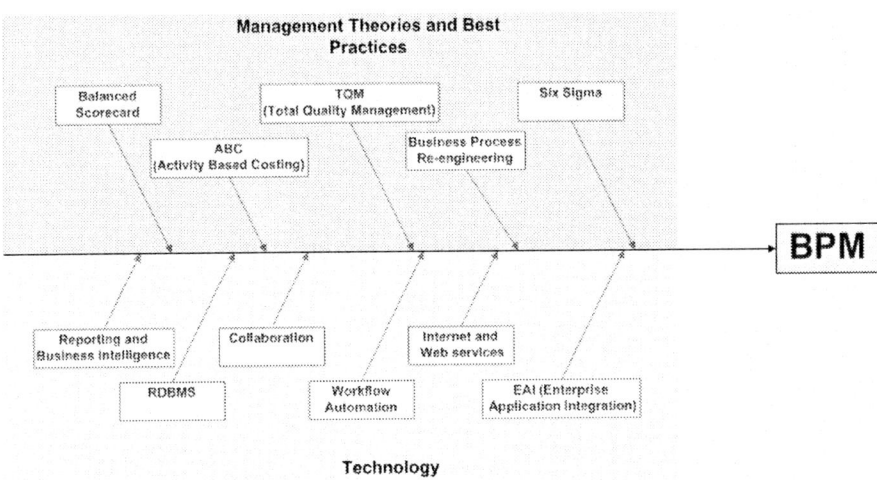

Figure 3.1 Business Process Management – brings the previously standalone technologies under one roof and supports all existing management best practices.

Basically up until now we did not have the technology to digitally manage the end-to-end business process in real time. Everything was done via after the fact data. We now have the tools needed to manage the end-to-end business processes. But I will re-iterate that BPM is not just about technology alone. It is about a new way of working around the process. BPM is a new capability which will be embedded within the organization culture and within the IT infrastructure to attain process excellence.

> **Business Process Management (BPM) is the term applied to describe the capability and technology which enables organizations to model, automate, manage and optimize the business process leveraging existing IT infrastructure.**

In "Business Process Management- the Third Wave" the author writes *"Every CEO, CIO, CFO, supply chain director and management consultant on the planet has imagined business models and the associated business processes that they would like to implement right away. They are not short on new ways of imagining how to improve their businesses. Until the third wave every new business process has been hard to achieve in practice due to the cost, time and technical effort needed to implement the required software."* Although BPM brings many preexisting technologies under one roof, I will write about the main ones as below.

- Workflow Automation
- EAI (Enterprise Application Integration)
- Collaboration

Workflow Automation

Workflow has been around for a long time and there is a lot of confusion regarding workflow automation and BPM. Workflow is defined as a sequence of activities performed by participants to reach a common goal. Based on flow defined the workflow engine electronically routes the data and content required to perform each activity to a process participant. Workflow automation in its own right has the following limitations

Monitor/Optimize Processes:
Workflow automation does not provide the tools needed to analyze, monitor and optimize the business processes. Reporting tools which are provided by workflow automation are not sufficient to handle exceptions and optimize business processes. Workflow automation lacks the process management perspective.

Integration:
In a workflow solution automated activities like calling other software applications or third party products are implemented by the workflow engines API. That creates "tight coupling" architecture and defeats the purpose of agile and flexible business process. Workflow solutions do not implement EAI best practices.

To summarize workflow automation is about making routing of work faster and efficient, assuming that human tasks are the bottleneck. Routing of human work is one component of BPM. BPM focuses on managing the business process life cycle to make processes agile, flexible, visible, and efficient. BPM takes workflow automation to a new higher level. Workflow automation is just one component of BPM. We can safely say BPM is a superset of workflow automation. Table 3.1 shows the difference between workflow automation standalone and workflow automation in BPM.

Capability	Workflow automation	BPM
Focus	Task Routing	Process Management
Integration	Tight Coupling between Integrated Applications using custom API supplied by workflow vendor and Third party products.	EAI and B2B integration using open standards like HTTP, XML and Web Services.
Scope	Application Specific	Enterprise Wide and multiple processes
Process Modeling	Limited	Advanced Process Modeling with essential features like Collaborative Process Design and Process Simulation
Reporting	After the fact reporting	Real Time process metric reports and dashboard views.

Table 3.1 Difference between Workflow standalone and Workflow in BPM

Enterprise Application Integration

Business processes in the extended enterprise touch several applications, cross department boundaries and touch several partners and suppliers. Organizations have built heterogeneous systems over the years. It's hard to imagine a complex business process that does not get information from an array of internal and external IT applications. BPM implements EAI best practices to leverage information locked in other systems around a business process.

EAI was designed to bring together the software systems from disparate departments within an organization and across an organization by sharing data in an automated fashion. Many BPM products integrate with existing industry EAI products like Tibco, BizTalk, etc. Below are the differences between EAI and EAI in BPM.

Capability	EAI	BPM
Orchestration	Orchestrates data synchronization across heterogeneous systems within and outside the enterprise.	BPM orchestrates the business process involving both human and IT application participants.
Duration	Simple and short running flows of Data	Business Processes could be very complex and long running.

Table 3.2 Difference between EAI and BPM.

EAI is based on a messaging architecture. Heterogeneous applications can talk to each other using messaging and data adaptors. Changes in one system can be propagated and transformed to all other systems. BPM leverages the data integration capabilities of EAI and makes data integration work around the business process. BPM product must provide flexible integration capabilities to leverage EAI products. B2B (Business to Business Integration) using open standards like XML and HTTP and web services.

Business-to-Business (B2B) integration:
Addresses the interactions, collaborations and integration between business processes in separate enterprises. Communication between trading partners could be as simple as sending a standard message and expecting some action in return. B2B integration now has some standards like ebXML, RosettaNet etc. Another way to foster B2B integration is through Web Services. BPM product must support standard B2B collaboration protocols. Chapter 8 is dedicated to EAI in BPM.

Collaboration

Collaboration among the participants of the business process is an extremely critical element of BPM. Collaboration in the extended enterprise could be human-to-human or B2B (business-to- business).

Human-to-Human collaboration
Human participants need to share documents, have discussions, and need to collaborate around the process. Complex and long running processes can never be 100% automated. Complex and long running processes require human intervention. Human participants need to collaborate around the business process or around a particular activity of a process. Advanced BPM products must provide different kind of collaboration like instant messaging and discussion groups around the process among process participant's .The intelligence and business value generated as a result of collaboration must be persisted and made available for view at any

point in time. The business intelligence generated out of collaboration can be analyzed for process improvements. Chapter 9 covers collaboration in detail.

> Each component namely workflow automation, EAI and collaboration has managed one piece of the business in the past. BPM sets the axis of convergence of the above components and combined with new BPM tools and process methodologies provide powerful capabilities to manage end-to-end business process.

Working without a process mindset using BPM tools

The BPM team needs to be BPM savvy and must have needed training on BPM terminologies and technology. In the absence of BPM knowledge, BPM teams risk developing a workflow system without the "process management" focus. It would be synonymous with buying a Hummer and using it for grocery shopping. Many BPM teams do just that because they do not know what BPM is about. That is the reason I put a lot of emphasis on first getting the correct process mindset. They just develop a workflow solution, automate hands off of work and think they are process enabled.

In the absence of BPM knowledge, that's what we did when I initially started out. We got bogged down with BPM tool features and small details and missed the big "process management" picture. The business users continue working in their old fashioned way – only replacing their inbox trays with digitized queues. They continued to use the phone and email for process intelligence instead of process monitoring. Task got piled up for people on vacation, because managers did not use BPM built in work and resource management features. Again the point is getting the process mindset and getting to know all features of BPM is critical to unleash the full potential of this powerful way of working.

Business Process Management Systems (BPMS)
Here I will cover BPMS from a higher level just to put everything in perspective from a methodology and best practice standpoint

Business Process Life Cycle
Business processes are ever changing and have life cycle of their own. Figure 3.2 shows the business process life cycle. Process design, deployment, monitoring, exception handling, analysis and optimization of the business process are what constitute process life cycle management.

> **Business processes mutate, grow, shrink, split into sub-processes, converge and are constantly changing reflecting the ever changing Business Landscape.**

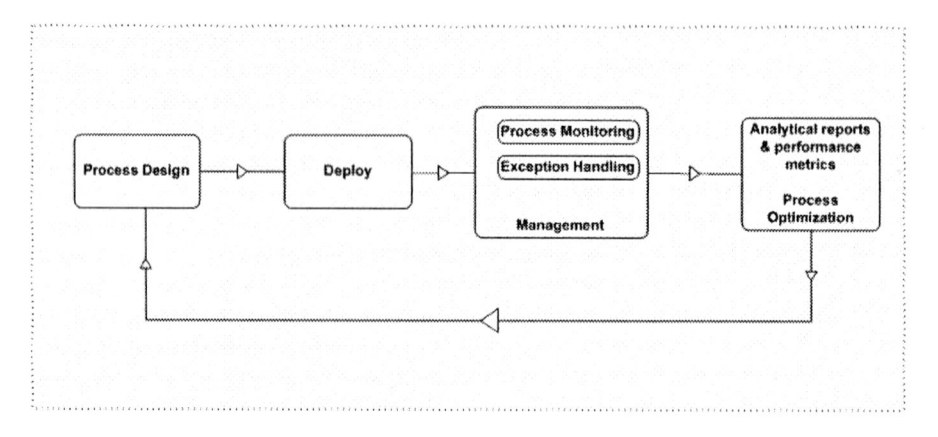

Figure 3.2   Business Process Life Cycle

Process Design:
    In the simplest terms a process model depicts graphically and describes a business process. It helps an organization understand the business process. A process design can reveal many process problems, bottlenecks and inefficiencies of the underlying process that appears to be functioning well. A process design is the first critical step in the process management lifecycle. Business processes are designed with a graphical process design tools called the ***process designer.*** Process designer is a integral part of a BPM product as shown in figure 3.3. The outcome of process design is called a business process diagram which is stored in a process repository.
    Collaborative process design is about multiple people working on process design concurrently. Once a design is completed the BPM team should test it by a technique called process simulation. ***Process simulation*** mimics a real business process in runtime in a controlled environment (time and participants). Simulation is extremely time consuming to do, but many mistakes are discovered here. Potential disasters in process design can be avoided if process simulation is learnt and practiced by the BPM team.

Process Deployment:
    Processes are deployed to the BPM server (process server/engine) from the process repository. That means that the business process is now live and users can execute it. As seen in Figure 3.4 the deployed process now resides in the "production database or repository".

Chapter 3: Business Process Management Systems

Process Monitoring/Exception Handling:
Business managers monitor the business process in real time using "process management tools". The objective of this phase is twofold

- 360 degree visibility of the process in real time to gauge the health of the process and in turn the enterprise.
- Handle any exceptions that might arise in the live running processes.

Process Optimization:
During this phase the processes are analyzed for increasing efficiency and better meeting the business objective. The input is derived from below factors

- The results of the Process Monitoring/Exception Handling Phase are analyzed for any improvement and fixing of errors in the original design
- Business Owners can factor in any business change in the optimization phase.
- Bottlenecks in the process which is causing the process to choke.
- Analyze Performance Metrics of running processes.

BPM Infrastructure

The new process centric BPM infrastructure will be built using existing IT infrastructure and new BPM technologies that abstract the business process into a process layer. Business Process Management System (BPMS) will enable process life cycle management using various BPM tools. The new process centric BPM infrastructure will be built on top of existing IT infrastructure like existing databases, user directories, enterprise applications, portals, etc.
Figure 3.3 depicts a concept diagram of BPMS.

> **BPM is the fabric that glues all the existing IT infrastructure and new BPM tools together to create a process centric IT infrastructure. BPM leverages not replaces existing IT infrastructure.**

Work or Task Portal

A work portal is the end user environment where users can access their work lists. Think of it as "You have got work" notification area. BPM products typically have a work portal to facilitate human participants to view, complete, return, reject, and reassign their work items. Advanced BPM products also enable corporations to integrate with existing industry portals like plumtree, etc. This capability by providing a single point entry within existing portals makes people comfortable as they view BPM as an extension of their portal way of working. Fig 3.3 shows enterprise wide users accessing the business processes and work lists through a portal.

Chapter 3: Business Process Management Systems 42

Conceptual Model of BPM Infrastructure

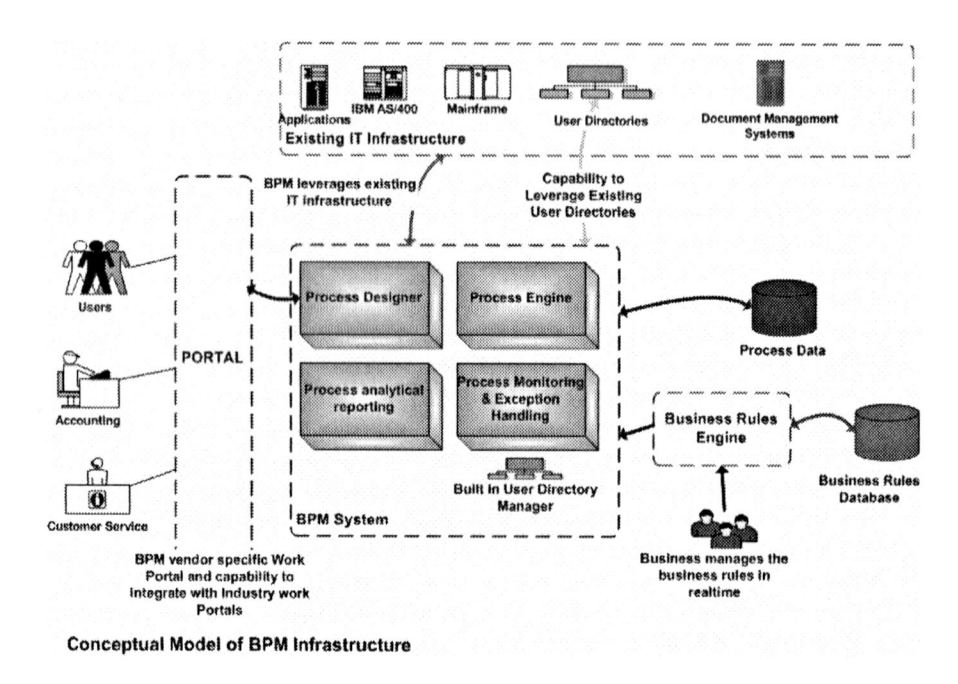

Figure 3.3 Logical/Conceptual diagram which depicts an enterprise BPM system.

User Directories

BPM products should have the capability to allow organizations to define and manage their organization charts i.e. the roles and relationships (like team structures, departments, etc) within the BPM system. Most large organizations have digitally stored their organization structures. BPM products must also allow organizations to leverage existing user directories like LDAP, Active Directory, JNDI, etc. by providing seamless integration. Fig 3.3 shows the BPM system being able to maintain user directories internally and at the same time integrating with existing user directories. Work is routed to humans based on user directory information.

Business Rules Engine

For certain organizations in highly volatile industries like banking and wireless where product offering change constantly, it is important to abstract the business rules into a business rules layer. Most BPM teams get confused between business rules and business processes and tend to link the two. Due to this confusion the process design suffers as business rules which are constantly changing are embedded into business processes.

Chapter 3: Business Process Management Systems 43

> **Business Rule:**
> A business rule is guidance that there is an obligation concerning conduct, action, practice, or procedure within a particular activity or sphere.
>
> **Business Process:**
> A business process is a group of related activities that when executed in a particular order together creates value for a customer.

Some BPM products have BRE (business rules engine) built in. Separating your business processes and business rules is prudent if your business operate in a highly volatile industry. Embedding business rules in business processes cripples the business processes. Most business rules engine support structured English like notations and can be managed by the business users. Some examples of business rules are

- Pricing structure at a large grocery chain.
- Wal-mart changes pricing on products on a weekly basis.
- Government and State Regulations. (Sales tax calculation for a worldwide customers)
- Highly Volatile Industry – like Banking.
- Pricing and billing for Telecom industry.

If your BPM product does not come with a built in BRE and you are in a volatile industry there are many third party products available which can be linked to the business processes as shown in figure 3.3. The advantage of having a business rules engine in a highly volatile industry is

- Empower the business owners to manage the business rules in real time.
- Business rules can be reused by multiple business processes and other IT applications thus lowering maintenance cost and eliminating the need for rules silos throughout the organization.
- Consistent execution of business policy by encapsulating all the rules in a central location.
- Efficiency gains by moving the business rules into a central location with streamlined management of the business rules.

Enterprise Application Integration Capabilities
The BPM server must have enterprise strength EAI capabilities as the process centric infrastructure will be built on top of existing infrastructure. BPM does not invent the wheel as far as EAI is concerned. BPM uses the same EAI concepts and can integrate with existing EAI tools, the only difference being in BPM the BPM server will orchestrate the EAI activities based on conditions described in the

Chapter 3: Business Process Management Systems 44

business process design. As shown in figure 3.3 BPM product must have robust EAI capabilities and must have the following

- Integration with existing EAI products like BizTalk, tibco etc.
- Ability to discover and invoke web services.
- Ability to integrate with existing document management systems (DMS). Documents play a critical role in digitizing business processes.
- Ability to connect to various data sources using universal ODBC and JDBC connectivity.
- BPM product must have script like features to integrate and invoke existing Windows and Java based applications.
- Ability to participate in B2B collaboration and transaction over the internet. (see figure 3.4)

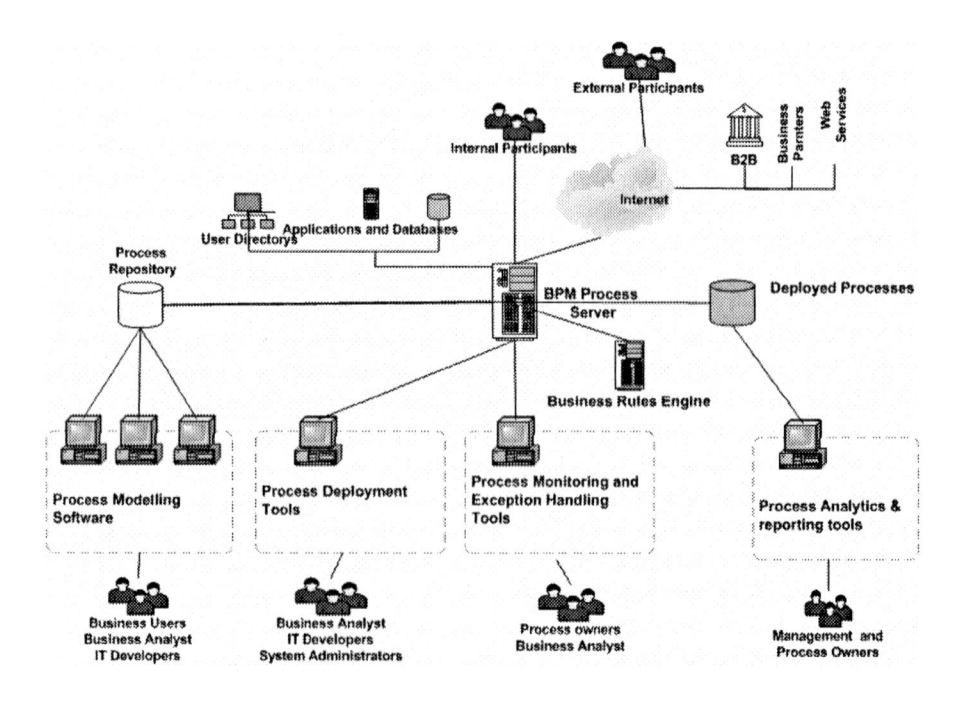

Figure 3.4 Typical BPM physical architecture.

Figure 3.4 depicts a simple BPM architecture. Enterprise BPM infrastructure will be a function of various factors like BPM product, existing IT infrastructure, complexity of the business, transaction volume, number of partners and suppliers involved, etc. The BPM physical architecture diagram is just to paint a picture of a typical BPM architecture for first time BPM implementers. The BPM server executes processes, has EAI capabilities to connect to existing applications, can communicate to partners, suppliers and customers over the internet using HTTP, XML and Web Services. In summary the BPM server orchestrates the business process at the same time provide process visibility and control to the business mangers. The business managers have the power to stop orchestration, change the path of orchestration, restart orchestration, etc in real time.

Benefits of BPM

Howard Smith and Peter Fingar authors of *Business Process Management – The Third Wave* say "By acquiring BPMS now companies gain unprecedented control over the management of their business processes, supplementing their existing systems and accelerating the achievements of business objectives." Below is a summary of the benefits of BPMS.

Agility	The Process is at the Kernel of BPMS .The Foundation of Business Process Lifecycle Management (Design-Deploy-Manage-Analyze/Optimize) is that business process will continue to change. Management of business process change is the objective of BPMS. BPM infrastructure and tools will allow Existing Applications, Partners and Suppliers to act as Process Participants using EAI and Collaboration thereby creating a "Plug-and-Play extended enterprise". This kind of Process Centric IT infrastructure will be extremely Agile and will be an enabler of change.
Visibility	Process Management Tools enable Business Owners and Managers to view and manipulate Live business Data from a Process context. Event Notification services create process visibility.
Flexibility	Processes and business rules can be changed in real time. Exceptions are managed in real time. BPMS are built on top of existing Applications and infrastructure.
Efficiency	BPMS increase the speed of executing a Business Process by workflow automation, Collaboration, Proactive Alerts and Notification – thereby reducing error and Manual tasks

Chapter 4:

Business Process Design

Chapter 4: Business Process Design 47

Companies tend to think of themselves in terms of management structures and departments, but customers see the organization in terms of the value it delivers to them – it is the business processes that deliver this value. Every company should view itself in terms of the business processes it executes. By doing so every company can improve the way it delivers value to customers, and hence profitability and success of the company itself. The first step in acquiring BPMS is process discovery and design. Before the BPM team starts process design it is critical for the BPM team, stakeholders and senior management to

- Understand and document how business strategy is linked to the particular business process.
- Define the problem that needs to be addressed. The goals of the process design efforts should be clear. Depending on the type of business process the primary design goal and secondary design goals should be established. It could be visibility, efficiency, cost reduction, etc.
- Develop key performance metrics against which the process can be benchmarked once it's deployed.

Most organizations have problems in discovering the process requirements. The first step in business process design is to discover the process requirements. This is the most important phase and many process projects fail to deliver the expected rate of return because sufficient time is not spent on process design and simulation. From my experience the cost to fix a major process design error once it has been deployed is enormous. Figure 4.1 shows some of the problems associated with deploying a bad process design.

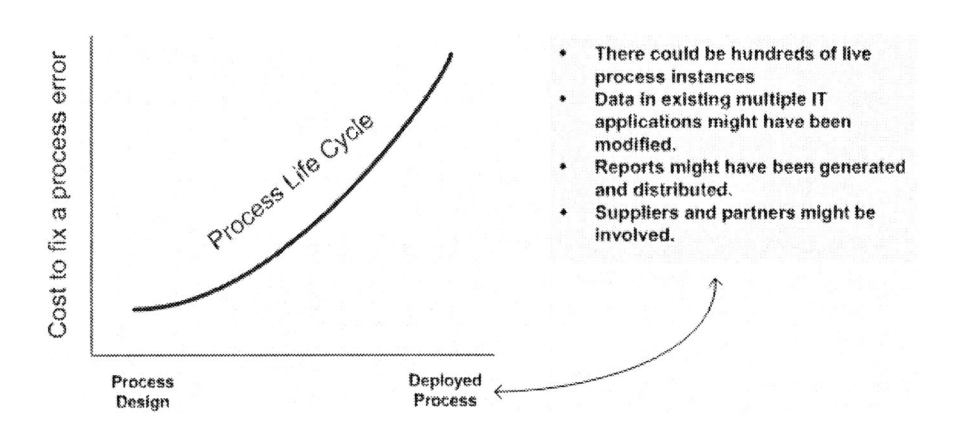

Figure 4.1 Cost to fix a error in a deployed process could be very high.

From a customer and business standpoint below are some of the reasons why having a good process design is a critical success factor in enterprise wide BPM deployment

- Business processes are outward facing – meaning that is how the customers view the enterprise. Major process design flaws directly affect the customer and can hinder enterprise wide BPM deployment.
- BPM provides visibility – so any process design issue is easily visible creating issues for stakeholder, process owners and the entire BPM initiative.
- Hundreds of process instances could be live before an issue is discovered which makes taking corrective action time consuming.

But what does "business process design" mean? What is a "BPD (Business process diagram)"? Before I get into process design I feel the need to define what a business process is. Putting it simply "a business process is a set of logically related business activities that deliver something of value (e.g. products, services or information) to a customer". Once the BPM team knows the process requirements the next stage to use the graphical process designer to design the process.

> **Business process discovery and design is the method by which the BPM team understands and defines the business activities that constitute the end-to-end business process. The objective of process design is to ensure that the designed business process is effective, flexible, optimized and meets customer requirements.**

BPM Team

Business process design is the first step in being a process managed company. A top notch BPM savvy team is critical to the success of enterprise wide BPM adoption. An effective BPM team is made up of people from different departments. Top management support for the BPM team is a critical success factor.

The BPM team will be responsible for process life cycle management. The BPM team size and structure will be a function of the process complexity, organization size and industry. There is no one set of rules so I will briefly write about a typical team.

BPM Project Sponsor:
Usually the senior management who understand the value of being a process managed company. Project sponsor has a clear understanding of how business strategy is linked to business process.

Process Steward/Owner:
The process owner is that member of the organization's top management team who
- Is appointed by the executive team and given the responsibility, resources and authority to deliver and manage the business process.
- Provides overall direction for the BPM project.
- Participates, co-ordinates, manages and guides in the process design phase. This also involves arranging for the business process users and functional managers to get BPM training and creating process design teams.
- Plans and manages process development and deployment with the BPM team and the business users.
- Monitors business processes in real time for their use and performance.
- Continues to participate and manage in process optimization.

Business Process Analyst:
Is BPM savvy, has a process mindset and is familiar and comfortable with process life cycle and BPM methodologies. Business process analyst also knows the business fairly well
- Works with the business users and process owners to get the requirements.
- Documents all aspects of the business process for the BPM team.
- Participates in business process design including process flows, business rules and organization chart issues.
- Interacts with IT for automated activities and BPM infrastructure issues.
- Monitors business processes and participates in exception management.

Software Developers:
They work with business analyst to develop the automated activities.
- Work with the business analyst to understand the as-is and to-be process.
- Document and design the automated activities.
- Responsible for automated activities and BPM infrastructure issues.

IT Infrastructure:
- Maintain and monitor existing IT infrastructure and BPMS.
- Responsible for infrastructure availability and support.
- Maintains BPM infrastructure.

Process discovery and design

As we have seen business rules and processes are everywhere. Most manual or semi automated processes are **opaque** as they have evolved overtime and they are not well defined or documented. Some ways of working are explicit and others are implicit. I was in a BPM conference talking to companies who have implemented BPM systems. Most people I spoke to say the single biggest problem they had was

Chapter 4: Business Process Design   50

discovering their business processes. Most BPM teams spend a lot of time in process discovery. I am not surprised by that, and this time spent is a clear reflection of **opaque and rigid** processes most enterprises have.

To implement a successful BPM system, developing a robust business process design which reflects a optimized version of the real business process is critical. Figure 4.2 shows the methodology needed to develop a correct BPD. The BPM team (IT, business analyst, business owners and other business people) need to first discover the existing process. In the **process discovery** phase all details about the current way of working, who does the work, which IT systems are involved, where the work is done, any partners or suppliers, why it is done that way, what are the outputs of the process, etc are documented to form a picture of the **"as-is"** process. The **as-is process** represent the current picture of the process.

At this stage the BPM must look at the as-is process and identify areas that need to be improved. KPIs (see Chapter 7) are key performance indicators which need to be identified, so that when the process is deployed the success of the process deployment can be benchmarked. The BPM team must then use graphical process design tool to develop the **"to be"** business process design. From a higher level below are the some of the typical goals for the process discovery and process design phase

Process Discovery

- The output of process discovery is to come up with the "as-is" process.
- Understand the problems and focus on areas that need to be improved.
- Keep the final consumer of the process in mind. Consumer of process could be customer, partner or internal customer. What are the consumer's perspectives?
- What are the other stakeholder's perspectives?
- Value of business process – how it effects the organization.
- List areas that need to be improved or automated in order of business priority.

Process Design

- The out is the "to-be" business process design.
- Keep KPIs in mind.
- Leverage the power of parallel processing.
- For complex long running processes – create multiple sub-processes.
- For grey activities build in exception handlers.
- Separate business rules into a business rules engine if you are operating in a high volatile industry.
- Have loose coupling with existing and new IT systems.

Chapter 4: Business Process Design   51

Process Discovery and Design

Who? How? When? Data? Where? → Process Discovery Phase → AS IS Process → KPI Roles Data and Systems → Process Design Phase → To Be Process

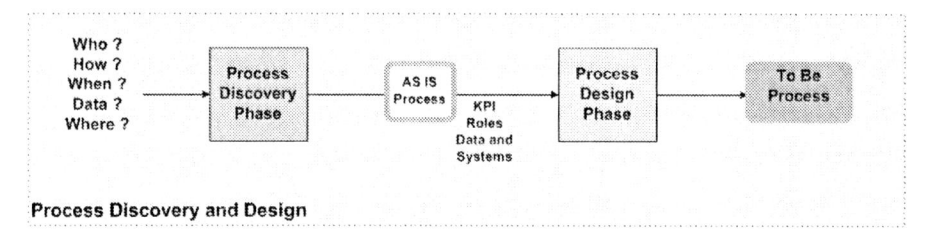

Figure 4.2 Process discovery and process design.

BPM Terminologies

The output of the design phase is a streamlined, comprehensive and a graphical view of the business process called the BPD (Business Process Diagram). Business process diagram depicts in a graphical fashion all the components of a business process namely process flow, activities, sub processes, participants, routing rules, events, etc. Having a robust and solid graphical process designer is extremely important. The tool should offer expressiveness to model complex business processes. The BPM team should also posses strong process design skills.

> **Purpose of this chapter**
> - **Help you evaluate a BPM product which has a top class graphical process designer which is intuitive, flexible and powerful to design enterprise strength business processes.**
> - **Help you design agile, effective and flexible business processes.**

By knowing that certain process design features exist the BPM team will be able to leverage them for process design. In the absence of such features or not knowing about them, the BPM team might end up creating less efficient business process designs or simply creating workflow automation. The whole idea of BPM is to bridge the business and IT divide. The process designer acts as a common language between the two. I want to make the following BPM terminologies clear as they will be used throughout the book

- BPD (Business Process Diagram)
- Process Instance
- Process Activities
- Process Participants
- Work Items (and Work Lists)
- Sub Processes

BPD (Business Process Diagram)

Business process diagram is the output of a business process design effort shown in a graphical form. The BPD is deployed to the process engine (BPM server) after testing. The process engine orchestrates the business process diagram. Today most BPM products use different notations and semantics for business process diagrams. Most BPM vendors store business process diagrams in different proprietary formats today. The Business Process Management Initiative (BPMI) is a non-profit initiative that has developed a standard Business Process Modeling Notation (BPMN) [www.bpmi.org]. The goal of BPMI is to standardize process design so that the process design diagram becomes BPM vendor independent.

Process Instance

A business process is a set of logically related activities performed in series or parallel to deliver something of value (e.g. product, service, information etc) to a customer. When the business process has been deployed and is being orchestrated by the process engine (BPM server) – the running live instance is called a **process instance** and is identified by a process identifier (think of it as case id) to uniquely identify and differentiate it from other instances. The running process instance has a status like active, completed, waiting, stalled or suspended.

Active	Process instance is being orchestrated by BPM server and is running.
Completed	Process orchestration has completed and the process instance has reached the end step.
Waiting	The process instance is waiting for an external event.
Stalled	The process instance has come to a standstill due to an exception.
Suspended	The process instance has been temporarily suspended by the administrator. The process instance can be resumed later.

Process Participants

Participants are the entities who perform the various activities in the business process. Participants could be humans, IT applications or sub-processes. The BPM server generates work items for the human participants. Human participants access these work items through a company portal or a BPM portal. The BPM server integrates and consolidates the non human participants in the process orchestration. System participants are invoked by the BPM server.

Process Activities

Process activities can be human or automated. Human activities require human intervention like data entry or approval. Automated activities are activities which do not require manual intervention like generating reports, sending out emails or calling other IT applications. The activities have a life cycle from activate to complete. Different BPM vendors might use different specific status terms, but generally speaking activities have active, late, complete, return, aborted and failed status representing different phases of the activity life cycle.

Active	Status indicates the activity is active and a work item has been generated if it is a human activity. It is residing in a work queue waiting for a participant to pull it out and work on it or it is in the work item list of a participant (s).
Late	Activity is late and most likely an action has been taken, like supervisor has been notified.
Complete	Activity has completed successfully.
Return	The activity was returned to the sender for further information or comments. A human participant can return a work item if participant thinks the previous participant has not completed the actively in whole.
Aborted	Activity has aborted and could not complete in specified time.
Failed	Activity has failed due to an infrastructure issue.

Work Items (and Work Lists)
  When a process is initiated work items (task items) are generated in response to activities and process rules and distributed to the respective participants. The human participants can view their work items using a portal. The work list is a collection of work items from one or more process instances. The work list can be accessed by the human participants in a variety of ways like through integration with the company intranet, company portal, custom application or Web.

Sub Processes
  When a parent process (main process) calls another process – the called process is called the sub-process. Different BPM vendors might call it sub-process, child process or nested process but the idea is the same. The caller process is the parent process. The call could be synchronous or asynchronous.

  Let's bring all above terms together with the help of an example. Figure 4.3 shows the customer support process of a XYZ software company. The customer can either submit a product enhancement request or initiate a customer service request. Customer can initiate the customer support process via the web, email or over the phone. The customer will need to enter in customer account number and

- Based on the customer request type the process takes the customer support path or product enhancement path.
- The process figures out the type of customer (individual or corporate) and proceeds accordingly.

Chapter 4: Business Process Design   54

- If it is a corporate support request the "Update CRM" automated activity is triggered which updates an existing CRM application. After that a work item is generated for a human participant. When the issue is resolved "Notify Customer" automated activity sends an email to the customer.

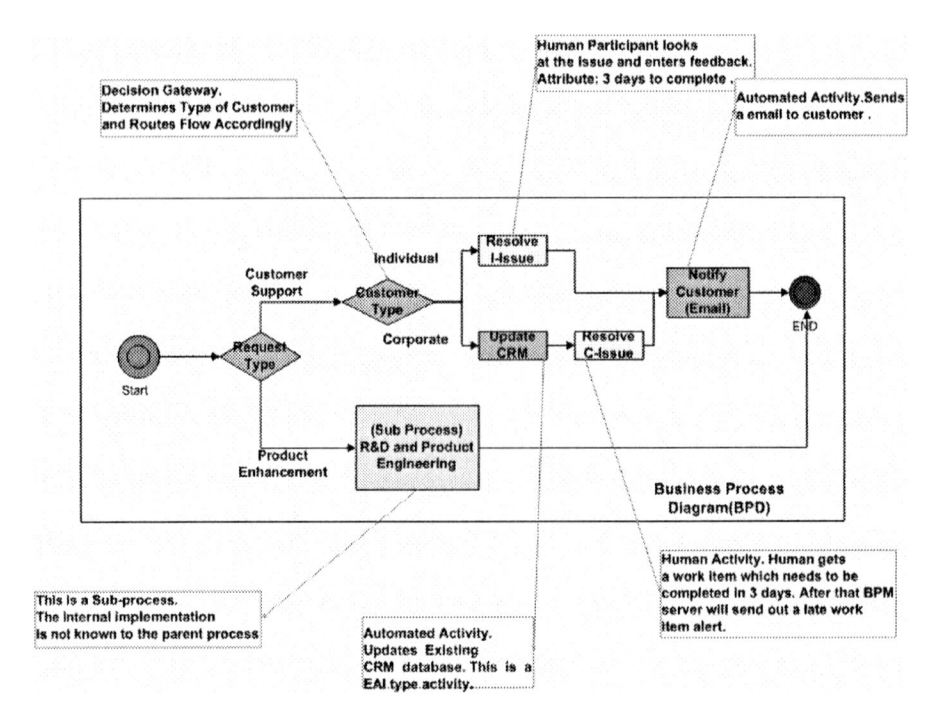

Figure 4.3 Represents a business process diagram for a customer support process

A client initiates a product enhancement request as shown in Figure 4.4. The request type decision gateway checks for the request type. Since it is a product enhancement request the sub-process "R&D and Product Engineering" is triggered. The dark black line shows the path of execution. This running process instance is identified by a Process ID.

Let's take another example, an individual initiates a customer support request because he has been having problems with the software. Figure 4.5 represents another process instance with a different path of execution. Figure 4.4 and 4.5 represent two process instances.

Chapter 4: Business Process Design 55

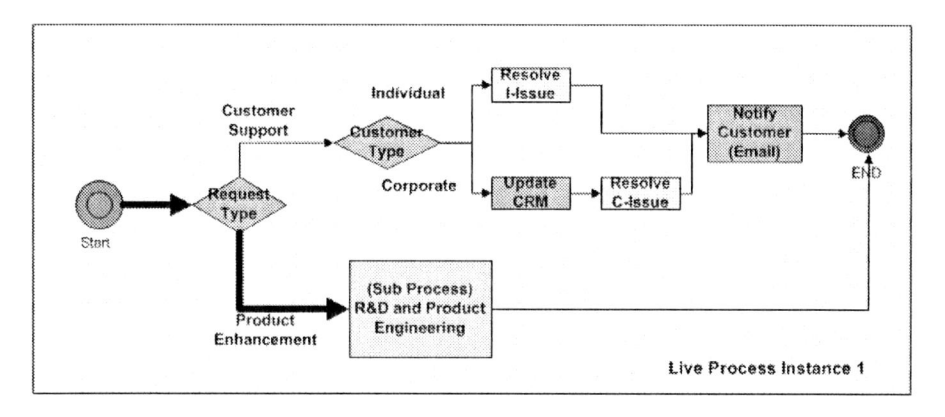

Figure 4.4 Represents a running instance of the Customer Support Process.

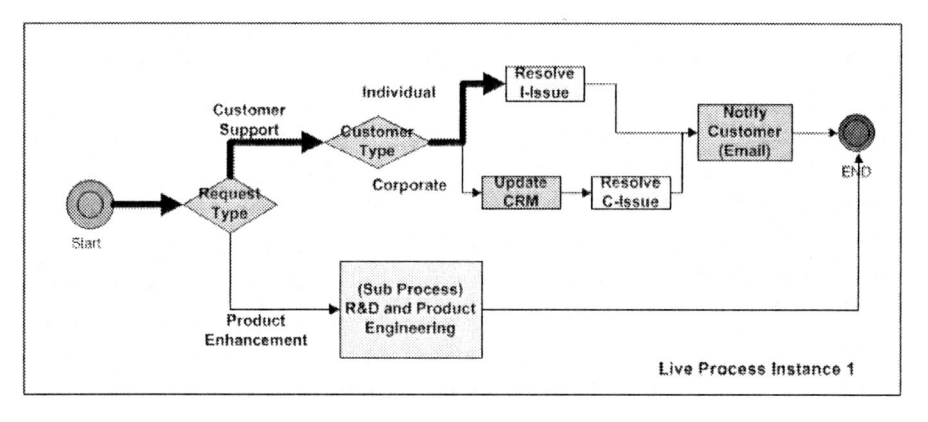

Figure 4.5 Represent another running instance of the Customer Support Process.

BPM Tool Requirement – The Process Designer

The process designer is the common language between the IT team and the business team. Even if the IT people need to be involved to do hardcore coding for the automated activities, the process design diagram is still easily interpreted by the business users. So with BPM we now have this common ground whereby the business user and the IT team can evaluate what they are doing and how useful it is.

Because real world business processes are complex, the graphical process designer must be able to support the design of sophisticated processes without the need for complex programming. The process designer must be user friendly with point and click, drag and drop features to enable the business people to design

Chapter 4: Business Process Design 56

business processes. Because business processes require combining both human and system participants the graphical process designer must be able to reflect and integrate both systems and people. The process designer must provide mechanisms for the BPM team to specify any kind of automated activity and have existing applications be participants of the business process. Human participants need to make decisions, enter information and handle exceptions and errors.

> **Process designer must be a single tool that can be used by different audience- business and IT for designing the business process. Process designer should have a visual design environment that is tightly integrated with the execution environment so that the deployment actually means running the process design.**

A good process designer tool must be flexible and powerful to cater to all process stakeholders. A graphical process designer is an integral component of a BPM product. The key components of a robust process designer capable of modeling complex business processes are

- Intuitive process flow design
- Multiple process repository support
- EAI activities
- Multiple Form Support
- User directories and routing
- Nested Model Support (Sub Processes)
- Collaborative Design
- Process Simulation
- Event Notification Services
- Process documentation

Intuitive process flow design
The process designer must be rich and powerful to easily model real world complex processes. Process designer must be readily understandable and easy & intuitive to use by the non-technical people like business process analyst or process owners. Process designer must enable the design team to embed automated activities as part of the process design.

Multiple process repository support
BPM design team would like to create multiple process design repositories. BPM product must support multiple process design repositories. Enterprise processes can then be grouped and stored in the appropriate databases. This is important as processes related by function, value, security or service can be grouped and stored together. Process design segregation enables companies to modularize architecture. Also IT infrastructure activities like back up, maintenance, etc become

Chapter 4: Business Process Design   57

smooth with pockets of downtime. Process designer must have features to easily transfer business process diagrams between different process repositories.

EAI activities

Embedding various EAI activities in the process design in a point and click manner. Process designer must have built in integration capabilities, support for web services and interface to existing EAI products. The process designer must allow at design time to map process data transfers to EAI activities in a point and click manner.

Multiple Form Support

Human participants need a user interface for participating in the business process. Work items are generated by the BPM server and assigned to human participants. Human participants enter information and review information through the form interface.

- The process designer should out of the box provide a built in form designer. The form designer will be used to create forms for human activities. The ability to easily design (like WYSIWYG- what you see is what you get editor) electronic forms is critical.
- Some companies would like to stick with their existing portals and forms for consistency sake and also because the business users are comfortable with the existing forms and portal. The process designer must provide the capability to support electronic forms in any language like ASP, HTML, InfoPath, Word, Adobe, JSP, Excel, etc.

User Directories and routing

Most large organizations have complex organization charts which are constantly changing. The BPM product and in turn the process designer must have the capability to leverage existing organization charts. The BPM product must also allow the creation of new organization charts within the BPMS. The BPM product must also have features which enable process designer to create **virtual queues and groups**. You can think of a queue as a shared in-box.

The BPM team then has the flexibility to design business processes which can use existing organization charts or organization charts created within BPMS. This is important because the existing organization charts might not be efficient from a process standpoint. The process designer must be able to provide the BPM design team the flexibility to route human activities to human participants using multiple ways as described below. Figure 4.6 shows an organization chart for a typical organization with CEO at the top. The different division VPs report to the CEO. The VPs manage different departments like sales, marketing and engineering. Each of these departments has several employees. The BPM design team must be able to design work item routing using the below methods:

Chapter 4: Business Process Design   58

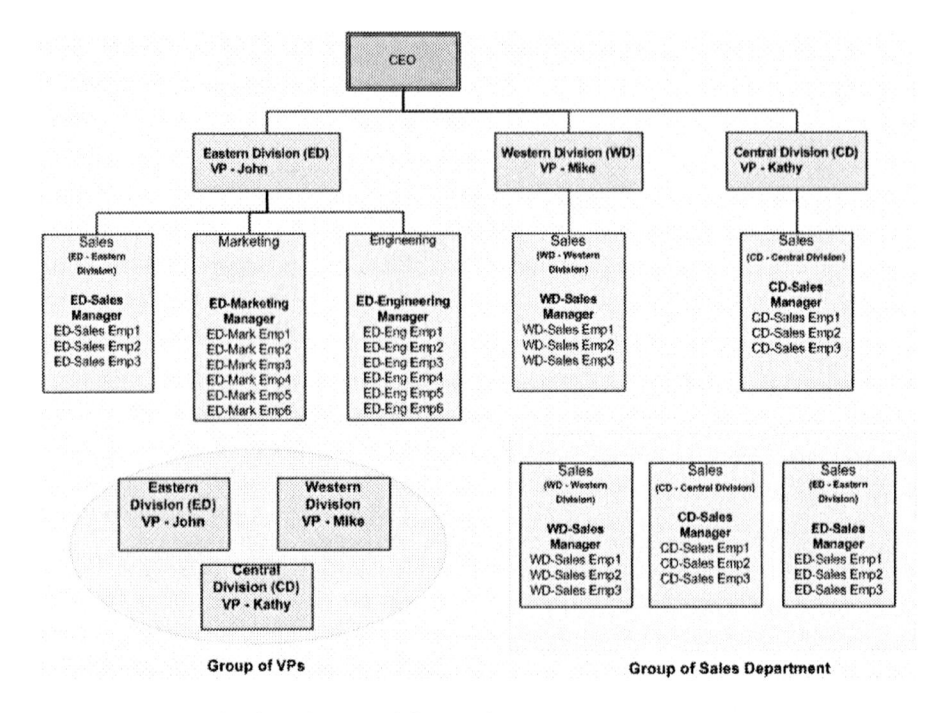

Figure 4.6 Organization chart used for routing

Name Based Routing
Work items are routed by names using this type of work routing. Name based routing is not the best way to route work items for complex business processes. The design team has to hardcode human participants name in the business process diagram. If the employee leaves the organization the process design will have to be modified and redeployed.

Ad Hoc Routing
Allows BPM design team to dynamically set the human participants of the work items based on certain business rules or data in existing systems. These business rules could be residing in existing systems and could be determined at runtime using automated activities. For example in a sales process, based on the account number entered by the client, an automated activity could determine the account manager and route the work to the respective account manager. That means the human participant names are not known in process design phase and are determined during the course of process orchestration.

Department based routing
Allows the BPM design teams to route work items based on department names. All members of that department/function group will receive a particular work item in their work list. For example from figure 4.6 a particular report could be sent to all employees of the eastern division sales department namely "Sales ED (Eastern Division)".

Relationship based routing
This type of routing enables process designer to route work items based on organization chart relationships. So for example in the purchase order process a request to purchase is initiated by any employee of the sales department. The approval step will go to the initiators supervisor as defined in the organization chart.

Group based Routing
This type of routing allows work items to be routed based on groups. For example in figure 4.5 the BPM team has created a virtual group name "group of VPs" with all the vice presidents as members. The BPM team can use this group as a recipient for human activities.

Work Queues
You can think of a queue as a shared digital in-box. The BPM design team can create new work queues and specify which members of the organization chart have access to this queue. Work items can then be sent to this virtual queue. Members of the queue can checkout work items from the queue, view the work item and either work on the work item or return in back to the queue.

Multiple Organization Chart Support
The ability of a business process design to be linked to multiple organization charts is very powerful and can create process efficiencies. Also a process activity can be linked to more than one organization chart based on certain conditions. For example based on the time of customer support request
- A person in Asia could get a work item based on the Asian Org Chart or
- A person in North America could get a work item based on North American Org Chart.

This is a very powerful feature as companies can create multiple organization structures for efficiencies and offers customers seamless and consistent service.

Nested model support (sub-processes)
In the extended enterprise business processes are long running transactions cutting across departments, customers, partners and suppliers. For process efficiencies and design efficiencies the graphical process designer must have capability to invoke other processes. The invoked process is called a sub-process and the calling process is the parent process. The parent process is able to pass information to the sub-process and receive information from the sub-process. Basically the sub-process becomes a participant of the parent process. From a business standpoint this feature is extremely important for the following reasons

- Collaborative design so that different people/companies can working on sub-processes and main processes.
- Business process re-use.
- Business process outsourcing.
- Complex processes can be broken into simple sub-processes.

The parent process should be able to call the sub-process in a synchronous or an asynchronous fashion. In a synchronous communication the parent process waits for the sub-process to complete before proceeding. In an asynchronous communication the parent process fires off the child sub process and continues.

Collaborative process design
Designing complex business processes is a team effort and requires various kinds of specialized skills. Therefore the process designer must support collaborative process design. Collaborative process design allows multiple people located across the globe to work and collaborate on the same business process design concurrently without any kind of conflict. This team effort is called collaborative process design.

Collaborative design enhances the speed of the process design effort as multiple people can work on different activities at the same time. The quality of the process design is better as experts from various disciplines (IT, User Interface, Business Analysts etc) can work on the activities they specialize in. To support collaborative design process designer must have robust **process versioning** capabilities.

Multiple members of the BPM design team will need to work on different activities of the same business process. To prevent work overrides and to manage the entire collaborative work the graphical process designer must provide some kind of process versioning mechanism. BPM design teams must be able to check out entire processes or individual activities of a process and work on them. Basically the features must mirror the kind of version control we have in tools like VSS (Visual Source Safe) for software development life cycle.

The graphical process design tool must allow the re-use of human or automated activities by storing them as object or components.

Chapter 4: Business Process Design

Process Simulation

One way to validate a business process design prior to deployment is by process simulation. Process designer must have process simulation capabilities. In the simplest terms simulation is running the process in a controlled business environment to validate and test it. In a compressed period of time and space simulation is one way to validate a process before releasing it to the users. The following are the potential benefits of process simulation

- Non-Value-Add steps can be identified and reduced.
- Bottlenecks can be identified and removed.
- Process Design meets expectations.

With a new business process it's virtually impossible to guarantee whether the process will work as expected and asserted ROI will ever be realized. Simulation provides a mechanism for business process design validation under simulated business conditions and can substantially reduce the risk of deploying a faulty new process.

Simulation is quite difficult to do, it is time consuming and most design teams do not do it. Also learning how to design simulations models will take time and experience. Simulation involves projecting or reproducing the behavior of a business process. A model for simulation is the data and events in a controlled time period. Simulation models for business process also depend on the nature of the underlying business process. Below are two examples of different kind of business processes and the simulation problems that arise

(a) *A physical process system*, for e.g. Internet ordering process or supply chain process. Let us consider the internet ordering process. In such a kind of order fulfillment system the objective might be to speed up the delivery of a customer order and allow a customer to track the status of the order. The process designers can create several simulation models based on several permutations and combinations of location of customer, location of distribution center and type of product. As you can see simulation models can become extremely complex, difficult to manage (due to the number) and time consuming.

(b) *A management process system*: For e.g. A Automobile Insurance Claim process. Here the end value product for the customer is Claim settlement dollar amount so that the customer can get his car fixed. From start to finish several entities could be touched like finance, legal, rental division, adjusters, police, Registry of Motor Vehicles etc. The Information generated from the process is consumed by different entities in different forms like reports, letters, faxes and data feed to other systems for decision making. So creating simulation models for this knowledge process would require careful planning.

Event Notification Services

Event notification services of the BPM server create the event driven enterprise. As the BPM server orchestrates the business process and participants do their work, the BPM server raises events which can be captured. Alerts and notifications are an integral part of BPM systems and provide event information to relevant people.

> **The process orchestrator (BPM server) knows when an activity is late or completed or is in danger of being missed. Appropriate events are fired to notify the relevant people to minimize any disruption.**

Alert is a message signaling immediate action and notification is a publish kind of message like "You have a work item in your inbox". Business process events are fired by the BPM system as the business process is being orchestrated. Alerts and notifications are post event processing to signal something happened. Task arrival, late tasks, late processes, automated task error, etc are examples of some events that fired. The graphical process designer should provide capability to declaratively create alerts and notifications with customized messages. Declaratively means with out any programming in a point and click fashion. The following are some of the most common type of alerts and notifications a BPM system would generate.

- New work item arrival.
- Updated business process available.
- New business process available.
- Completion time of a live process instance has expired.
- Completion time of a work item has expired.
- A work item has been is completed.

Customized alerts with process data in the following events

- Process instance has been stalled (come to standstill).
- Activity has been aborted.
- Activity Failed.

Process documentation:

The process designer should have the capability to generate process documentation using the BPD as the source. Generating business process documentation using the graphical process designer is extremely important for the following reasons

- Business processes change constantly. Manually keeping process documentation up to date is extremely expensive – rather I would say it is virtually impossible.
- Compliance requirements like – global rules, federal, state, ISO 9000, CMM, etc.
- Process documentation can act as specifications for new partners and suppliers.
- Process documentation can be used by the business process QA team to test the business process.

Business process design best practices

I cannot cover business process design in detail here – but based on my experience (mistakes and lessons) over the past 3.5 years with BPM I want to mention couple of process design best practices. Below is a limited list of process design best practices.

Event Notifications (Alerts and Notifications)

These simple yet powerful BPM services will catapult your business process to high visibility category. BPM design team should leverage this feature which requires no programming. Alerts and notifications add **broadcasting capabilities** to the business process. The BPM team must use built in alerts and notifications wisely. Process owners can take proactive actions in case of problems. Alerts and notifications keep the customer in the loop. Below are some examples of Alerts and Notifications

- An order has been placed.
- An order has been cancelled.
- Customer Service task is late.

> Do not overuse alerts and notifications. Customers and senior managers hate to be bombarded with every little information. Know the events which are most important to the business. Provide them with an option to stop receiving notifications if they want.

Work in parallel

This design philosophy will make business processes efficient. In the business process diagram position human and automated activities for parallel processing if possible, thereby increasing the speed of execution. Unrelated activities can be designed for parallel processing. Some activities which require a lot of input will obviously need to be in their right place in the orchestration chain – but other activities which require very little work hands off and input can be candidates for parallel processing. For example a report to be generated requires independent input from three participants. These input gathering human activities can be placed in parallel to converge to report generation. The following kinds of notations can be used for parallel processing.

Split	And-type	Used to specify parallel outgoing paths of execution
Split	Or-type	Used for Conditional branching
Join	And-type	Used to specify joining parallel paths of execution to a single path of execution.
Join	Or-type	Used to join alternative path of execution and to connect loops.

Web services

For automated activities and EAI (Enterprise Application Integration) activities as far as possible use Web Services (for details see Chapter 8). Web Services are called using open standards in a loosely coupled fashion. Using web services it is now possible connect to heterogeneous systems in a cost effective manner. This makes the business process agile and flexible.

Make sure web services being orchestrated support versioning. In BPM different versions of the same processes could be running side-by-side. Hence it is necessary for web services to support all versions of running process instances (see chapter 5 – Process deployment for details).

Sub- Processes

For complex business processes the BPM team should try to split the business process into multiple sub-processes. The payoffs are numerous

- Collaborative design as different members of the team can work on different process. Collaborative design also speeds up the process design phase.
- Break chunks of related functionality into sub-processes. It makes the entire business process modular.

Chapter 4: Business Process Design 65

- Having process functionality split into logical sub-processes make the parent process easier to validate and test. The sub-process can be tested standalone for input, value it delivers and output.
- Sub-processes also make continuous process deployment easy. This will be clear when I discuss side-by-side process versioning and in-flight process change in Chapter 5.
- Sub-process help in process monitoring. Based on the issues and areas of improvement the relevant sub-processes can be monitored.
- Sub-process also makes benchmarking easy. If the parent process is not doing well because of one sub-process it is easy for the process owners to easily spot the problems areas and make improvements.
- Common enterprise wide functionality can be embedded in sub-processes and used by multiple parent processes. That enhances process re-use which has a positive effect on process consistency and quality.

Exception Handling

It is no longer possible to consider exception handling as a secondary process design issue or worse a mechanism added after the fact. Exception handling needs to be thought of and built in the process design phase, and BPM supports that. Exception handling is a necessity in today's dynamic environment. Chapter 6 covers Exception handling in detail, but at this point I would like to mention the below exception handling best practices

Most graphical designers enable BPM teams to trap events which the BPM server raises. The BPM team should take a proactive approach for exception management design. Remember below is just a partial list to get the concept of proactive exception management clear

- Notifications for late process instances and activities.
- Notifications for overdue activities.
- Notifications for new version of process available.
- Notifications for a returned work item.
- Alerts for failed and aborted activities.
- Alerts for failed process instances.

All above alerts provide clues to a generated exception or an exception in the making. The process design should be such that should an exception occur, there must be a fallback or recovery path. For complex and mission critical business process the BPM team should think about building "**dummy human intervention**" activities which will only be triggered if a certain type of fatal exception occurs. That way if an exception does occur the dummy activity provides a door for the process manager or owner to enter and make corrections and overrides to fix the problem. Another way is to build **negotiation** activities. A work item goes to a person who acts as a negotiator. He sends the work item to various stakeholders, aggregates feedback and completes the activity.

> **Activities are performed by resources (Human or system). Any resource could become unavailable at any point in time. Design your business process in such a fashion so that there is a fallback or recovery path.**

Cluttered business process design
I have seen companies design business processes with, lets say > 100 activities. The BPD looks cluttered with activities connected by lines. Exception management, monitoring and optimization become extremely difficult with a poor process design. Trust me simply putting all real life activities on the business process design map will not make you process enabled. Careful planning and design skills are needed. Use sub-processes and web services. The business process design should reflect an optimized version of your real life business process. This will be a first step, and continuous process improvement will make you a process managed enterprise.

> **Business process design should not be a digitized version of your "as-is" process mess.**

Interdependent Activities
BPM design team should try to make different activities self-sufficient and not interdependent on each other. This might be difficult in the early stages of learning, but as the BPM team gains experience in the process life cycle the advantages will be apparent. Altering one activity in the business process should not disrupt the business process at other points or downstream. Continuous process improvements will be smooth and in-flight process data change will be possible due to the relative independence of activities. Experimentation will be possible which will lead to further refinement of the business process.

Compensating Activities
If the business process updates several different databases through different applications it might be required that either all or none updates go through. Due to the long running nature of business processes, BPM servers do not provide distributed transaction support. This can be achieved by compensating activities which are triggered if an exception occurs and you would like to undo a previous update.

Chapter 5:

Business Process Deployment

Process deployment phase is a critical success factor in a successful BPM implementation. Designed business processes are validated and deployed to production. At this stage most organizations forget about the people aspect of BPM and tend to concentrate on technology implementation. BPM projects even with well designed business processes are stopped or slowed down, because the people who will be using the BPM system are not on the same page. The visible backlogs of overflowing in-trays are gone, only to be replaced by overflowing digitized inboxes and overflowing digital queues. The end result is significant culture resistance and reluctance to cooperate with further iterations.

BPM server acts as the central command center for orchestrating the end-to-end process after it has been deployed. Process participants can now initiate the deployed business processes. Most vendors, analysts, books and white papers talk about deploying business processes to the BPM system as easy as one click publish and can be done by the business people. That might be possible for very simple business processes with no automated activities. For complex processes that is wishful thinking. I have implemented enterprise strength BPM systems and my experience has been that process deployment requires deep BPM knowledge, planning and a deployment strategy. I have yet to see a complex business process which can be deployed that easily by making a change and clicking the deploy button.

> Enterprise wide success for BPM projects depends on people as much as on technology and this applies to the deployment phase as well. BPM teams should not forget about the people dimension of BPM. Moving from paper trail based manual or functionally semi-automated process to a digitized process managed enterprise using BPMS involves unlearning old ways of working and learning to work in an event driven enterprise.

From my experience first time process users must be given the time and training to learn how BPM is going to effect them. They need to be part of some kind of testing or pilot project which is done prior to deployment. Organizations can plan for a BPM pilot to get the people accustomed to the BPM terminologies and BPM way of working. Once the business users are comfortable with the BPM style of working, successive pilot projects will not be needed. Pilot projects will help the users become BPM savvy, have a process mindset and become comfortable with BPM style of working. Trust me BPM style of working is pretty different than the way business users work with IT applications today. In the BPM style work comes to you, work is monitored, work is reassigned in real-time, work is returned, work is forwarded and exceptions are handled on the fly.

If the business users who are going to be using the BPM system are located across the globe in various offices – BPM process champions should be selected from various offices who will participate in the BPM pilot project. These BPM process champions will represent their respective offices and will gain

Chapter 5: Business Process Deployment

understanding of BPM and how work is done in BPM landscape. They should begin to appreciate the advantages of working with a process mindset in a BPM environment. BPM process champions will act as go to persons in their respective offices for any process or BPM related issues or questions. The key ingredient to BPM success is to involve the people in all phases.

Business processes need to be tested and validated in a non-production environment by the BPM team before pilot project. Proper non production environments like QA, pilot and staging need to be set up. Mistakes and erroneous design assumptions are discovered while testing in non-production environments. Figure 5.1 depicts a process deployment strategy that can be followed to be successful. There is no one style fits all strategy here, but whatever you do make sure people are involved and business processes are tested prior to deployment to production.

BPM Deployment

Step 1 Setup BPM QA and Pilot Environment	Step 2 BPM Pilot	Step 3 Process Deployment to Production
• QA and Pilot could be the same • Deploy process to QA • BPM team test the business processes • IT will test Business processes, automated Activities and entire BPM infrastructure	• Business Users become familiar with BPM way of working • Business Users test business processes for functionality	• Deployment of Business Processes to the BPM Server for orchestration

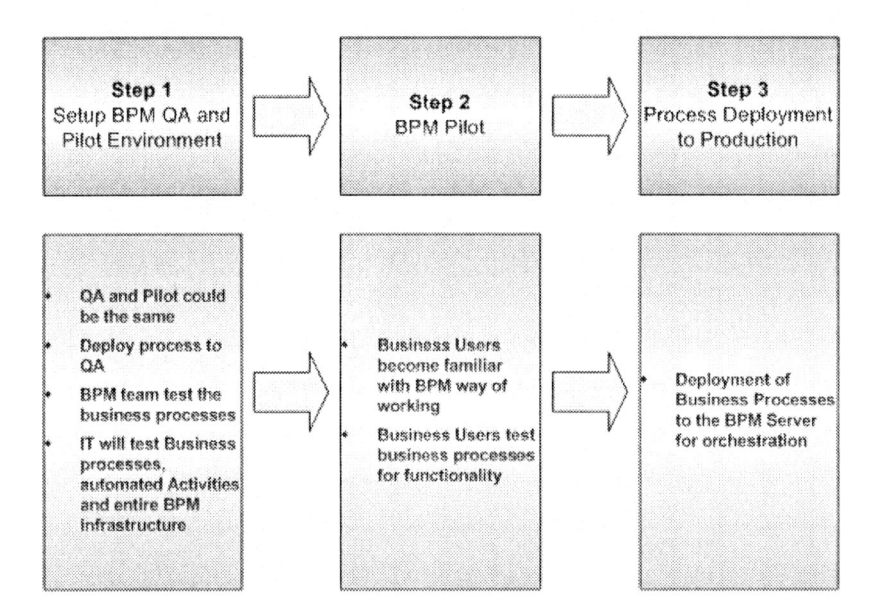

Figure 5.1 Process deployment strategy.

> **The BPM Pilot project deals with why BPM technology is here in the first place – to support the achievement of business objectives and business objectives are created and made possible by people. Pilot project also gives time to the business users to be comfortable with BPM and BPM style of working.**

BPM environments

As mentioned before BPM project cannot be managed as a traditional IT project, and as we shall see this applies to the testing and deployment phase as well. Large organizations might have different QA and pilot environments. In small organizations the BPM QA and pilot environments could be the same. The organization must have different environments for development, testing, pilot and production. Most large organizations will have the following environments as shown in Table 5.1.

Environment	Description
Development (Set up already. BPM team designs the processes here)	The BPM design team will work in a collaborative mode to facilitate business process design and simulation in this environment.
QA(Quality Assurance)	The BPM QA team will validate the business processes in this environment. Business processes can be long running and cut across departments, partners and suppliers. The BPM QA team needs a dedicated environment where they can create and leave long running processes instances. The BPM QA team might consist of QA members from partner and supplier firms as well.
Staging	Staging environment mirrors production environment. Since development and QA will always be in a state of change due to the process life cycle, staging can be used to reproduce, diagnose and troubleshoot production process problems if need be. Some firms do not have this BPM environment
Production Environment	Business processes are live and business users can work against them. The production BPM server orchestrates the business process.

Table 5.1 BPM environments

Process testing in QA

The BPM team must do the first round of testing in BPM QA environment before conducting a pilot. Business processes for the extended enterprise cut across departments, partners and suppliers so there must be a mechanism to test the external processes as well. Partner and supplier teams might need to get involved depending on the process and relationship. Partners and suppliers must provide QA/test environments against which their services can be tested. EBay provides a

Chapter 5: Business Process Deployment   71

test environment for vendors to test their processes. If that is not possible and there is no need for partners to get involved, the input and output results need to be validated when supplier/partner processes are called. This is called black box process testing where only input/outputs are validated.

IT people must validate all the automated activities and IT infrastructure. More details about IT infrastructure validation are given later in this book. Since BPM leverages existing IT infrastructure it is critical that supporting IT infrastructure is reliable and available. From my experience many process errors occur due to IT infrastructure issues. Figure 5.2 represents the difference between deploying a new business process and deploying an updated business process. When deploying an updated business process there might exist live process instances from the previous version.

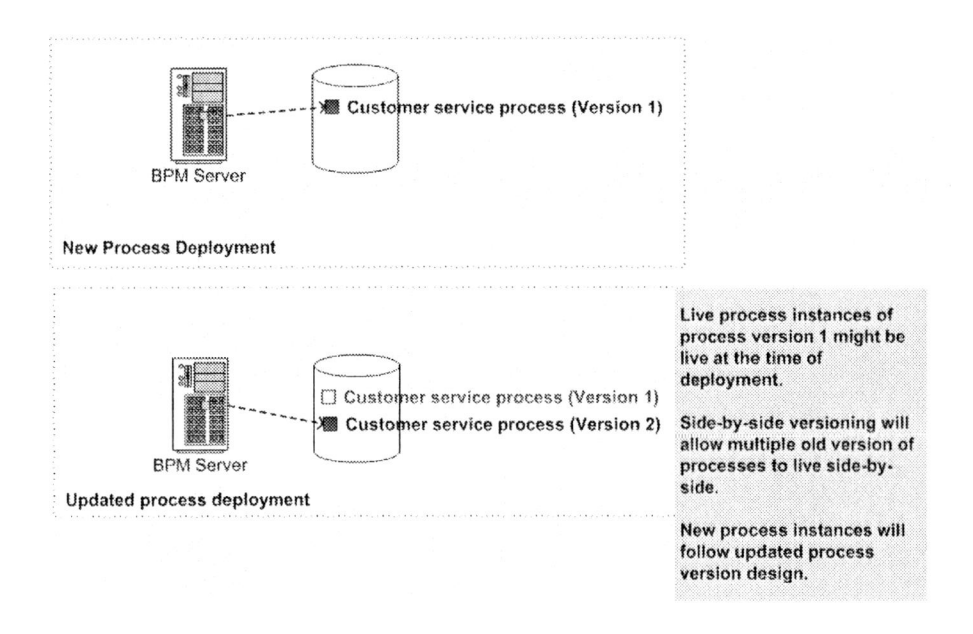

Figure 5.2 New process and updated process deployment.

New Process Deployment

For new business processes which are being deployed for the first time – process validation and testing will be a little easy from a business and technical standpoint as the BPM team does not deal existing process instances (discussed later). Test cases will vary widely and my aim is to provide a general guideline.

Table 5.2 describes a summary of the different types of test cases needed process validation.

Version 1.0 (First time deployment) process test cases	
Process Flow Paths	Validate all execution paths.
Sub-Processes	Validate all sub-processes are invoked in the correct fashion and validate input/output.
Alerts and Notification	Verify notification and event services are working as per design.
Automated Activities	Validate all automated activities. IT will work hand in hand with the business users on this one.
EAI	Validate BPM server is integrating and effectively orchestrating other IT applications.
Exception Handling	Validate if there is an exception, it can be managed and there is a way to continue process orchestration.
User Directory	Validates workload management.

Table 5.2 New Process deployment test cases

Process Flow Paths:

The BPM team should document and try to test for all possible process flow paths. Real world business processes are very complex and testing all possible permutations and combinations might not be possible. If the BPM team finds out there are a large number of possible flow paths – this might be a design problem. The BPM team should go back to the design board. The BPM team should think about using sub-processes to reduce the number of possible flow paths. Having sub-processes makes the parent end-to-end process modular, manageable and easy to test.

Sub Processes:

In the extended enterprise – multiple partners, suppliers, internal departments and customers can be participants of a business process. From a BPM context some of the above entities might serve the main process as sub-processes. Sub-process need to be invoked in an accurate fashion by the parent process. Correct, complete and real time information needs to be supplied to the partners, supplier and customer sub-processes. Depending on the sub-processes, these sub-processes might return information to the parent process. Basically the input into sub-processes and output back to the parent process needs to be validated. Sub-processes can be called in a synchronous or asynchronous fashion.

Chapter 5: Business Process Deployment   73

Alerts and Notification:
All notifications and alerts generated by the BPM system should go out to correct recipients, in a correct format and on time. Alerts and notifications are built by business users in a declarative manner. The BPM system manages and tracks these alerts and notifications. A complex business process will have several alerts and notifications and it might not be possible to test all. I recommend testing a few alerts and notifications just to make sure the BPM server works as advertised by the vendor. If the BPM server is robust and works as advertised the BPM team can safely depend on the BPM tool for functioning correctly.

Automated Activities:
Automated Activities are activities which do not require human intervention. Below are some examples of automated activities. Depending on the type of automation the required resources (mail servers, fax machines, shares, hardware, etc) need to be in place and available. The BPM team should then test all automated activities. Some examples of automated activities are

- EAI Activities (I will discuss testing these activities in a separate section)
- Generating documents with process data.
- Sending out customized emails.
- Distributing reports (Word, Excel, Adobe, etc)
- Sending a report to a FAX machine.
- Saving a generated document in the company Document management system.
- Sending a Voice Mail.
- Initiating other sub-processes.
- Manipulating and/or updating data.
- Writing customized log files.

EAI (Enterprise Application Integration) activities:
EAI activities form an integral part of BPM. EAI activities must be invoked in a loosely coupled manner from within the BPM process. The following test cases need to be prepared for EAI activities

- Correct data being passed to the web service or related IT application.
- Test whether EAI activities are working as expected.
- Test for unavailability of resources needed by EAI activities. What this means is all resources (databases, applications, etc.) which are directly or indirectly required by the EAI activities need to be tested for unavailability. The desired result is that the process orchestration must stop at the EAI activity when a needed resource is unavailable. The activity status must indicate failed or aborted. A BPM team member must be able to manually activate/restart the EAI activity when the underlying resource

becomes available. That way process orchestration can resume after underlying problem has been fixed. This is also an example of exception management.
- If the IT applications are enhanced then IT department must make sure that business processes that depend on it are not negatively affected.

Exception Handling Testing:

Chapter 6 covers exception handling in detail. From my experience exception handling testing is the most important part of testing a BPM team should do. No matter how good the process design is or how good your IT department is, exceptions are always going to happen. Having a strategy to deal with exceptions is extremely important. From my experience business spends 80% of their time in dealing with exceptions. In a BPM system the thing that annoys business managers the most is having a **stuck or stalled process instance or activity instance.**

A stalled process is one in which the live process instance is stuck and cannot continue. The results can be deadly like lost business opportunity, customer dissatisfaction, back office expense, etc. A good BPM product will allow modifying process data on live processes. So process owner can assign the work item to someone is available, restart the failed activity and get the process going.

> **Basically any resource (human or automated) could cause an exception at any point in time during process orchestration. Process design and Test cases should be such so that the process instance can recover from the exception.**

User Directory Testing:

In today's business climate organization structures are fluid –people move around, leave and join the company, new departments are created, existing departments are merged or purged etc. Organization Chart changes are going to happen and the BPM system must be robust enough to withstand any changes in real-time. The BPM team should prepare test cases for some scenarios mentioned below

- Employee leaves the company.
- Employee goes on vacation.
- Employees are moved around as a result of a re-organization.
- Organization creates a new department.
- Existing department is removed.
- Two departments are merged to create a new department.

This is not a complete list, depending on the complexity of the business process and organization structure test cases would need to be formulated.

Chapter 5: Business Process Deployment 75

Updated Process Deployment

For process deployments which are improved versions of existing process design, the BPM team must keep in mind that there might exist live process instances at the time of deployment. Most BPM products provide *side-by-side process version support* and *In-Flight process upgrade* features to facilitate smooth upgrade without effecting existing process instances.

> **Side-by-side versioning support:**
>
> New versions of a business process can be designed and deployed in production while old versions still have live running process instances. The BPM server will orchestrate the already running process instances according to their original version, while new process instances would run using the latest version.

> **In-Flight process updates is a very powerful feature which allows making process updates to existing process design and existing process instances. The process version number is not upgraded.**

Deployment Strategy for updated business processes

Depending on the complexity of the business process and process improvement made the BPM team needs to design a Deployment strategy as shown in Figure 5.3. As shown if minor changes are made then BPM team can decide to apply those improvements by updating the existing process using **in-flight process update feature**. This will ensure that the process and existing live process instances are updated. The version number will not be bumped up. Or they can decide to let existing process instances follow old process design and deploy a new version of the process. In this case the BPM team might need to perform side-by-side process validation (explained below) before deployment.

If major changes have been made (like existing automated activities, IT applications, process flow, etc.) then in-flight process update might not possible. Depending on the business process the BPM team would have to decide whether new version can be deployed with continuous operations or existing business process needs to be made unavailable for some time to let existing process instances complete (process downtime).

Chapter 5: Business Process Deployment 76

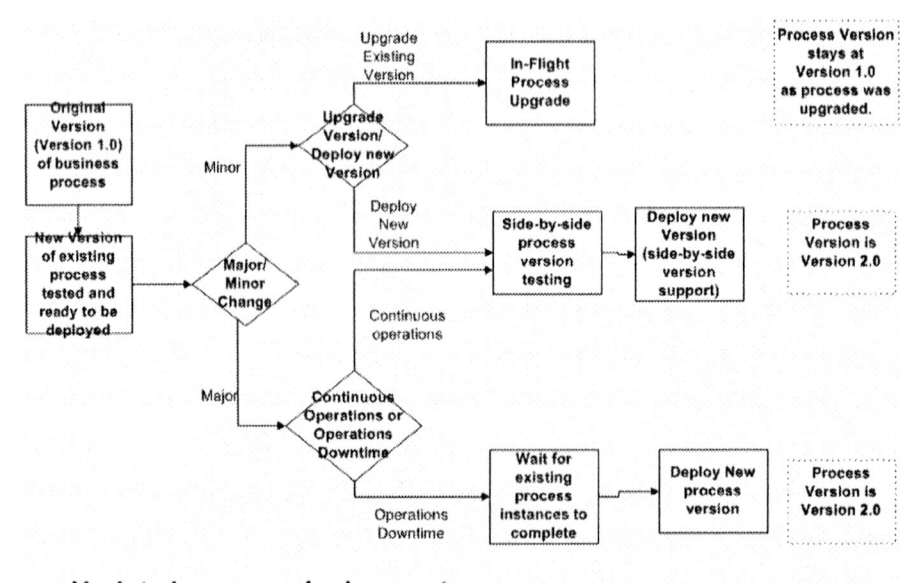

Updated process deployment strategy

Figure 5.3 Deployment strategy map for updated processes

Process Downtime (Deploy new process version after all existing process instances have cycled through)

The objective is to let all existing live process instances get completed before deployment. Once all existing process instances have cycled through the new process version can be deployed. This is achieved be the following steps

- Make the current process version unavailable in production. That is required so that no new process instances are created. The BPM team needs to co-ordinate with the process owner regarding the timing of this. Most BPM tools have this feature. Making a process unavailable does not have any negative effect on existing process instances.
- The BPM team should then wait for all existing process instances to be completed.
- Once all process instances are completed the updated version of the business process can be deployed.

The advantages of this kind of deployment are

- Less complex and less error prone.
- Less time consuming since side-by-side process validation does not have to be done.

The disadvantages are

- Downtime for the business operations.

This kind of deployment is possible if the organization has control over the business users of the process- like if the business process is used internally within the organization. An example would be "Expense Sheet Submission Process", which is an internal business process not exposed to external users. But for a business process like the customer support process where the customers are external to the organization this is kind of deployment is not possible. That takes us to the next option.

Deploy Updated Process with continuous business operations.
In this kind of process deployment the BPM team needs to ensure successful completing of existing process instances and at the same time make sure the new version of the process does not have any issues and will run successfully when it's deployed. This kind of deployment offers a transparent and seamless transition to the business users and ensures continuous business operations. The objective of this deployment is to

- All previous instances will run without any interruption and impact when new version of process is deployed.
- New process instances will be created using the latest process version and will run fine.

This kind of deployment is time consuming as side-by-side version testing needs to be done to support existing process instances.

Side-by-Side version testing
The objective is to make sure the new process version will not break existing process instances. So let's say the current process is at version 1.0. And let's assume there are several process instances live of version 1.0. Now when version 2.0 is deployed, all existing version 1.0 process instances must be able to complete successfully. Below is a step by step process to accomplish side-by-side process version testing:

- The BPM team must create multiple process instances in QA with process version 1.0 in different stages of execution.
- Deploy version 2.0 of the business process.
- Complete all version 1.0 process instances.
- Create and complete process instances of process version 2.0.

The BPM team must be able to complete all previous process instances and initiate and complete all new processes instances without conflict. Basically the BPM deployment must support process backward compatibility

BPM Pilot

BPM pilot is a critical success factor in an enterprise wide adoption of BPM. A BPM pilot project is when the real users get their hands on the BPM technology and learn BPM way of working. The business users who are part of the BPM pilot team might have been exposed to BPM as part of enterprise or department wide training. But it is here where they experience BPM. BPM pilot project is undertaken to demonstrate the effectiveness and capability of BPMS. The pilot project begins with a statement of goals based on the business process (or processes). Key stake holders need to be involved at some level in the BPM pilot project.

A successful BPM pilot project will demonstrate the power and flexibility of being a process managed enterprise and will help justify expanding the strategy enterprise wide. The business process selected for the BPM pilot should directly support the strategic goal of the enterprise. BPM teams make the mistake of selecting a business process which is more technology focused and less strategically focused. The pilot team learns the new BPM way of working which helps creates a process mindset. Some of the following BPM concepts are picked up

- Initiating business processes through the work portal.
- Work list and work items.
- Returning incomplete work to previous participant.
- Process monitoring and exception handling.
- Process data analysis.
- Assigning and reassigning tasks.
- Collaboration during the life of a process instance.
- Alerts and notifications.
- Process metrics.

A successful BPM pilot project will create BPM enthusiasm within the organization and will create the right conditions to be a process managed enterprise. A failed BPM pilot project will hinder spread of BPM to other departments and processes and will be called just another complex technology.

Process Deployment to Production

The designed business processes are deployed to Production BPM environment. For complex business processes IT department checks some of the following to make sure IT infrastructure is available and reliable

- Check user directories (organization chart), role relationships, etc to make sure all business users are listed in the correct places according to the position in the enterprise.
- All databases, IT Applications (CRM, ERP, Sales, etc) which are EAI participants in the Business Process are in valid state, connectivity has been established and are ready to exchange data with the business process.
- The Production BPM environment has been configured with proper security and permissions.
- If the business process involves external partners and suppliers co-ordinate with them regarding the logistics before the business process goes live.
- The BPM team is in place for managing the live process instances and responsibilities are clear.
- Performance & Availability of the BPM server have been tested.

The process is then deployed to the BPM system.

Chapter 6:

Business Process Monitoring and Exception Management

CEOs, CFOs, and almost all business managers want to see where they're in real time. There's incredible value in having knowledge about the state of your business processes in real-time. Whether you run the business process as an enterprise or consume the business process as a customer process visibility is a win-win proposition. Process visibility lets the consumers see what is happening to their order request or service request in real time. For the business managers and process owners process visibility enables to see how business operations are running, provides critical information to understand customer needs, assess organizational strengths and weaknesses, respond to market conditions and improve their business processes. Process visibility is key to optimizing business operations.

Business environment is changing at such a rapid pace that it is impossible to build the most optimized and fault proof business process in the first shot. While working on BPM projects I have experienced business rules and processes change under my feet just after process deployment to production. No matter how good the BPM design team is, it is impossible to design perfect business process as change is constant. What is needed is to have a mechanism to proactively monitor running business process in real-time to ensure operations are running effectively and manage problems before they arise and after they arise. Ongoing process management and exception management provide visibility and control over your processes. Exception management refers to taking corrective action when a business process deviates from its correct path or stops moving all together. Figure 6.1 shows the process management phase in business process lifecycle.

Business process monitoring today is done by business intelligence reporting and alerts and notifications. Business intelligence reports are an array of reports produced by IT for the management which depict how the business has been running. These are sub-optimal ways for process monitoring. Real time monitoring does exist in some of today's IT applications, but these monitoring systems do not provide the required end-to-end process information needed. Most of the existing IT applications send out alerts and notifications when certain data changes or when certain events are triggered in systems. These alerts and notifications are useful for the operations people but business users cannot derive much business value out of these notifications and alerts. Table 1.1 depicts the difference between the monitoring and exception management approaches in BPMS and today's cluster of applications.

Chapter 6: Business Process Monitoring and Exception Management 82

	Monitoring	**Corrective Measure**
Traditional Application	▪ After the fact reporting. ▪ Standalone Alerts and Notifications without complete process information.	After the fact bug fixing as part of Software Lifecycle which is time consuming.
BPM	▪ Real time process monitoring via Graphical Process Diagrams with complete process information. ▪ Dashboards depicting the health of a business process.	Exception Management (Real Time)

Table 1.1 Monitoring and Exception Management approach in BPM and today's IT applications.

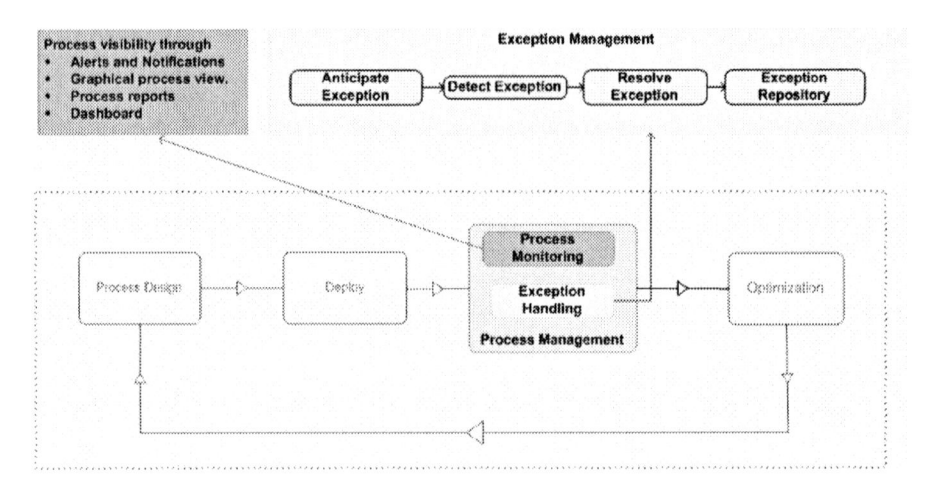

Figure 6.1 Process monitoring and exception management in a process managed enterprise.

Business Process Monitoring

Business process monitoring is required to make sure the business process in running effectively and is a prerequisite for continuous process improvement. You cannot improve a process which is opaque and cannot be measured. In order to improve processes, businesses must first understand where the bottlenecks are.

Chapter 6: Business Process Monitoring and Exception Management   83

> Without BPMS companies use sub-optimal techniques to monitor their operations which can be summed up as "Silo functional monitoring" and "Silo alerts and notification". Basically monitoring in most companies is done at the functional level.

Alerts and notifications
　　IT applications today have some kind of alerts and notifications built into them. These notifications are good for the operations people – but have little strategic value. Alert and notifications provide real time information about standalone activities, data change information or hardware problems. This information will make more sense if provided against the backdrop of the end-to-end process. Without the 360 degree process view it is impossible to derive any business intelligence out of the numerous alerts and notifications. As shown in Figure 6.2 below are the problems associated with monitoring today

- *Silo functional monitoring* where each department monitors their respective part of processes through after the fact reports generated.
- *Silo alerts and notifications* where department people get notified in case of issues or problems.

　　Basically monitoring is done on a functional basis. But business processes cut across department functions, so it would make sense to monitor at the process level. Here is an example why silo monitoring is not the optimal way to monitor and disadvantages.

Example
- Account manager of XYZ PC maker gets notified that a major customer has switched to a competitor. The account manager has no idea why this customer ended relationship all of a sudden. It turns out the customer was not satisfied with the customer service he was getting and had made several complains to the customer service department. Complains did show up in after the fact reports, but it was too late. Moreover other departments like Sales, Marketing and Accounting did not get this information.

　　Had XYZ PC maker been a process managed enterprise, the error of bad service complaints would have generated alerts and notifications to the highest level across the board, not just customer service manager. It would like ringing the fire bell that a major customer is dissatisfied and something needs to be done.

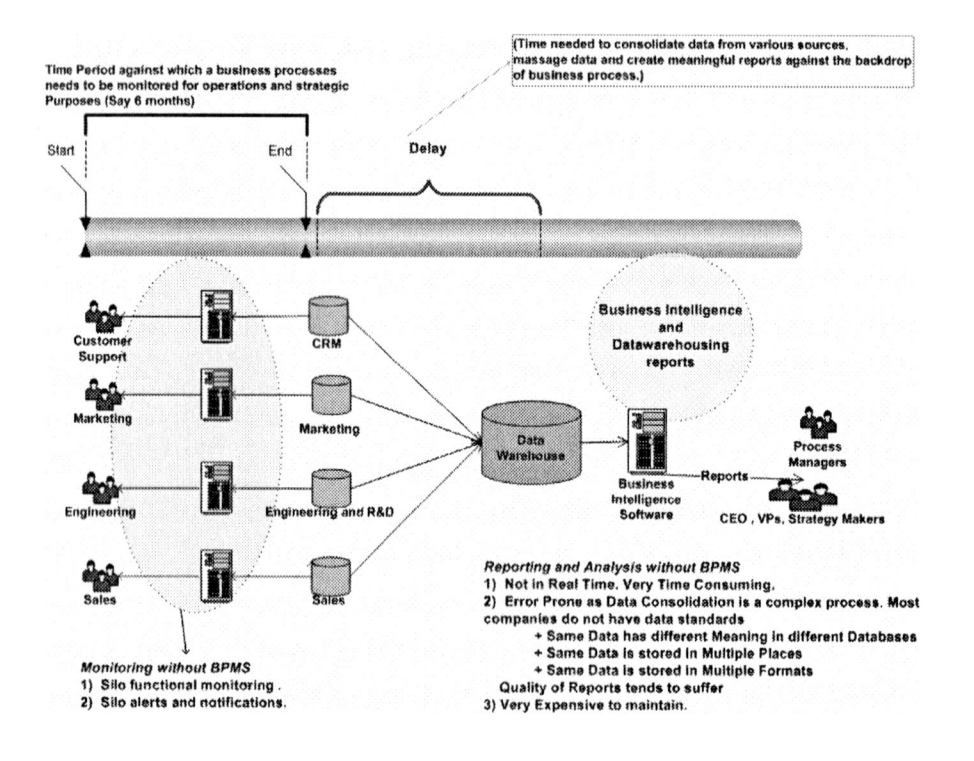

Figure 6.2 Disadvantages of traditional business monitoring.

In organizations which are not process managed, departments don't communicate. As a result there is very little cross monitoring. Turfs are protected without regard to customer. Cross monitoring would be losing turf and control. In BPMS information flows freely and process monitoring actually means cross monitoring. Sales can monitor R&D processes and vice versa.

> In BPMS alerts and notifications are sent to process stakeholders, not to departments. Alerts and notifications when sent at the right time, to the right people against the backdrop of a business process have strategic and business value and empower business managers to react proactively.

BI (Business intelligence)

Businesses have done a good job at collecting after the fact data in databases. Companies use this data to create complex reports for various purposes – predicting customer demand, customer analysis, launching new products, success rate of previous implemented strategy, cost of running processes, etc. Basically they are trying to (and they might not even know that) monitor business processes with the help of these reports. Data by itself has no business value. It needs to be correlated against business processes to derive any business knowledge out of it.

To create business value out of the terra bytes of data collected, companies have created huge data warehouses. Enterprise data warehouse have a significant **"delay period"** as data has to be consolidated from various data repositories into a data warehouse. The collection, transformation and massaging of data to create business intelligence reports of strategic value is an extremely complex, expensive, error prone and time consuming process. BI (Business intelligence) software is used to create business knowledge out of the aggregated data as shown in figure 6.2. Multi dimensional OLAP reporting tools are used to slice and dice this data. Common views are stored as OLAP cubes. Due to the delay factor these reports are after the fact. These reports are presented to department heads, senior management and boardroom for strategic decisions. Making strategic decisions with such kind of BI reports could result in bad decision, missed opportunity or even cripple a business.

Business intelligence in BPM

BPM products must provide analytic reporting capabilities on the process data generated by the BPM server. Some companies might want to merge process data in enterprise warehouses. Deep process metric analysis is done in the process optimization phase. I mention BI and OLAP reporting here because some of the reports might be common.

> **BPM products must include built-in OLAP reporting capabilities for analyzing process data.**
>
> **In addition process data structure format must be published, so that companies can integrate process data into enterprise warehouses for customized complex OLAP style reports.**

Figure 6.3 shows how business process monitoring in BPM falls in between real-time alerts and notifications and close to real time business intelligence reports. The philosophy of monitoring live and historical process instances, activity instances, workload activities, process times, activity times, delay times, etc creates real-time process visibility.

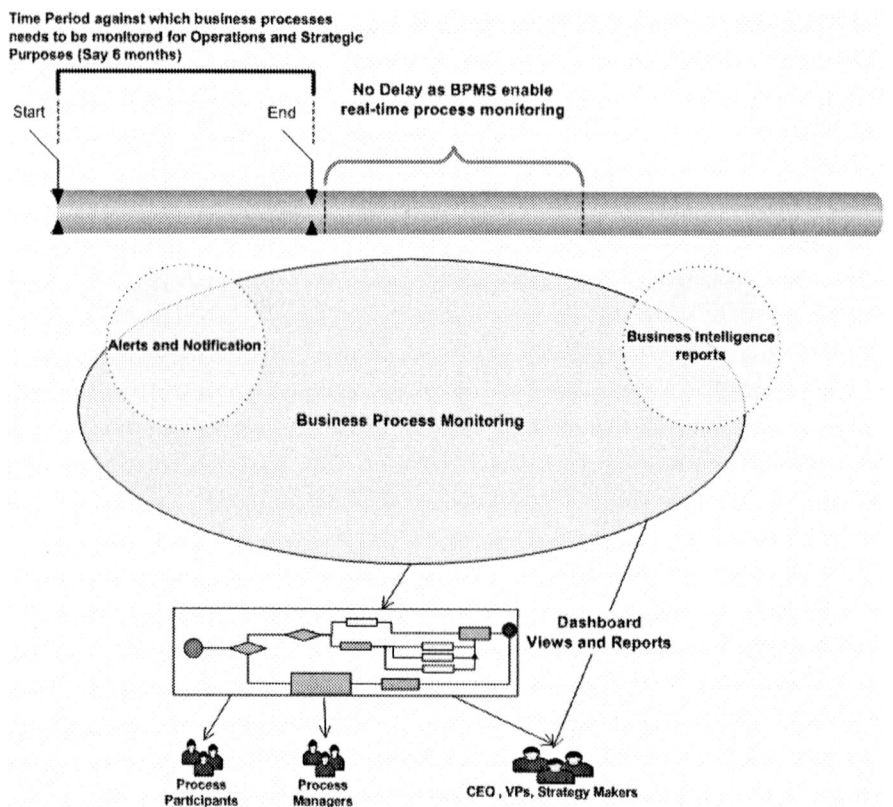

Fig 6.3 The place of business process monitoring

BPM product must have **process performance tools** which display process metrics in a graphical fashion using dashboards and process reports. A dashboard is nothing but simply a mechanism to display process metric data in a graphical fashion (charts, graphs, plots, etc) rather than in excel spreadsheets so that it's more intuitive. BPM performance tools must provide the core functionality to enable fast, easy and low-cost creation and reporting of an organization's process metrics. Below is a list of some of the process metric reports and dashboard views

- Process instances (live or historical) can be viewed graphically with all process information.
- Business users and managers can get live process instance views and process instance history reports (See Chapter 7 for details).
- Business users and managers can get live activity instance and activity instance history reports. (See Chapter 7 for details)

- Real time aggregation views and reports based on live process data and process history data. (See Chapter 7 for details)
- Real time workload management reports. Workload management includes, but is not limited to
    - Human activity reports
    - Workload activity reports
    - Late human activity repots

> **Process Monitoring requirements for BPM**
> - **Graphical views and dashboard enable process users to evaluate situations and take action based on real-time process data.**
> - **The graphical process monitoring view is the BPD (Business process design) in run mode. New symbols representing items like path executed and status of process and activities with complete information and participants.**
> - **Advanced BPM products can have features to graphically replay the process instance from the beginning explaining the execution steps that resulted in the current state or exception.**
> - **Ability to view all process and activity data as the participants saw it.**

> The appeal of business process monitoring which provides a graphical, clear and deep view into key business operations is understandable given the pace of today's fiercely competitive market, where real time reaction is key to operational efficiency and success. Business process monitoring in BPM with process centric decision tools and real time analytic computing power enable real time analysis and faster decision cycles.

Real time process visibility is helping companies to access critical business performance indicators in real time. This allows a company to know the pulse of existing business strategies. *Business process monitoring removes the smoke that exists in delayed analytical and after the fact reports.*

- Business process monitoring in BPM focuses on performance tracking, not deep analytics like BI. Deep process metric analysis is done in the **process optimization** phase. It positions itself as a middle layer between an organization's data sources and its applications, monitoring inputs and relaying only the important data items to the relevant people and/or systems.
- Unlike data warehouses the BPM system does not replicate or accumulate the massive volumes of raw data, resulting in a less complicated and expensive solution.

- Business process monitoring in BPM is proactive and deals with future-orientation, as opposed to BI which is reactive and addresses static, historical trends.
- Business process monitoring in BPM operates on a real-time event-driven model, while BI is time-delayed after the facts have been aggregated. Rather than depending on batch extraction and loading into a data warehouse, BPM systems trigger and update events as they happen.
- Business process monitoring in BPM uses rules-based monitoring and reporting. Reports are automatically generated on the fly by the rules that have been defined.
- Business process monitoring serves a much wider business user base, while BI is still the preserve of the back room for business analysts and power users and analysts.
- While BI is a standalone system, business process intelligence is closely integrated with operational systems and processes. It embeds intelligence deep into the business processes.

> **Business process monitoring takes business intelligence a new level - beyond strategic and tactical business applications to manage critical day-to-day operations.**

Exception Management

Without BPMS companies are always in **fire fighting mode.** That is the result of the rigid and brittle IT infrastructure created over the years. I have spent over 13 years in software industry, and my experience is IT spends 80% of resources in software maintenance. When business changes or there are issues with existing application **bugs** are logged, triaged, assigned and fixed.

Teams simply solve the problem at hand, without process insights. Teams implement minor improvements at best and never measure results. The symptoms go away, but underlying problems resurface in other parts of the process. Fire fighting efforts have no impact of process improvement. BPM will help companies move from fire fighting mode to exception management mode.

As we shall see exception management has a direct impact on process improvement. Businesses need to execute their business processes with high and predictable quality. Process quality is the fabric that will shine out. Poorly designed and ineffective business processes can kill a company. It is becoming extremely difficult to make complex end-to-end business processes **"Exception Proof"** due to

- Complex end-to-end business processes cross departments, partners and suppliers.
- Pace of business process change continues to be high.

Chapter 6: Business Process Monitoring and Exception Management

> **Exception is a deviation from the "optimal" (or agreed and acceptable) process execution path that prevents the delivery of services with the desired (or agreed) quality with maximum efficiency. Exceptions can create business disputes, generate strategy and financial risk, create significant back-office expense and can make damage relationships with the customer.**

This is a business process level notion of exception, where it is up to the stakeholders (process owners, process managers, process designer's and IT) to define what they consider as an exception, and hence as a problem they would like to address and avoid. Exceptions could be triggered by a variety of sources internal or external. Exception management is a BPM way of always being ready to manage any kind of exception. Some universities are doing R&D on exception management to bring it into the mainstream. The bug fixing approach is not going to work as the pace of change accelerates. Exception could be process, infrastructure or participant related as described below.

Business process exceptions

In the extended enterprise business process cut through heterogeneous applications, organizations, partners and suppliers. A poorly designed business process and not sufficiently tested process is one primary reason for exceptions. Reliable BPM infrastructure and reliable communication among participants are other reasons for such kind of exceptions. Here are some examples of exceptions

- Disruptive business events triggering exceptions.
- Unpredictable infrastructure failures across the process chain (partners, suppliers, etc.) and internet could cause process exceptions.
- Unpredictable BPM infrastructure failures can causes live process instances to come to a standstill unexpectedly. What that means is the BPM system is now unable to orchestrate this process instance and the process instance is stuck at a particular activity or state.
- Alerts and notifications getting delayed, lost or going out to the wrong recipients. Event notification services create the event driven enterprise. Alert and notification services are critical to operations and provide visibility to process users and managers.
- Automated activities like IT applications, web services, and automated scripts fail for a variety of software/hardware problems.
- Poorly designed business processes almost always results in various kind of errors
    - Process stalls due to bad process flow design.
    - Human activities are triggered without recipient data causing the process orchestration to go in limbo.
    - Sub-processes are invoked improperly.

Participant related exceptions

Process participants both human and automated have a commitment to perform their tasks as per their roles and contracts in the process. Any violation due to human error or otherwise results in participant related exceptions. Here are a few examples of participant related exceptions

- In Electronic commerce there is significant potential for fraud. For example a supplier over commits to supply Flat Screen monitors but fails to do so and disappears.
- A supplier provides bad quality information or products choking the process chain downstream.
- A human participant overlooks a prior bankruptcy filing for a friend and approves a mortgage loan for a million dollar house.
- A manager approves a $7000.00 (instead of $700.00) business expense. Human error among human participants is a major cause for exceptions.

It is impossible to make a complex business process **"Exception Proof"** due to the human dimension and infrastructure issues. Human beings are bound to make errors. Business processes are long lived transactions orchestrated by the BPM system involving an array of participants – human, IT applications, partners, suppliers, etc so the more the touch points the greater is the probability for exceptions. Also due to sudden business changes, a valid business can quickly be classified as an exception.

What is needed is a set of exception methodologies to proactively make visible, address and resolve exceptions. BPM systems provide services to manage exceptions. **"Exception Management"** is a key feature of BPM systems and managing exceptions is a way of working as shown in figure 6.4. Process owners and IT infrastructure people have to work hand in hand to proactively manage exceptions. Business process exceptions can be managed by the following methods

Exception Management

Anticipate Exception → Detect Exception → Resolve Exception → Exception Repository

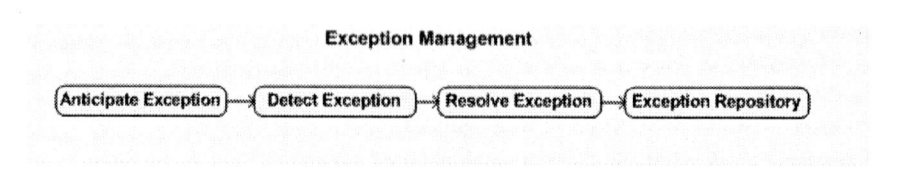

Figure 6.4 Exception management in BPM.

Anticipate/Avoid Exceptions

Monitoring the business process in real time will uncover many situations where an exception is likely to occur. Process visibility is a powerful feature in BPMS and process visibility helps process managers to avoid exceptions. A major auto insurance company has embedded a weather web service in its business processes. When there is a major weather warning in any part of the country relevant business managers are notified. For example the company can expect a flood of claim applications during the 36 inch snow storm prediction in the North East region of the country during the next few days. So business managers can proactively take steps to address the situation by having agents on call and/or hiring more contract people.

Detect Exceptions

Some exceptions are difficult to anticipate- but with business process monitoring it is possible to detect when an exception occurs in real time through alerts and notifications built into the process design and take corrective action. For example a customer service representative can proactively call a customer, inform the customer regarding an exception and negotiate a resolution.

- Due to supplier issue certain internet orders cannot be fulfilled. Customers are waiting for their orders to be delivered. Customer agents can detect this exception and proactively communicate with the effected customers on a possible resolution.

Exception resolution process

Once an exception has occurred the exception needs to be resolved. Most BPM tools provide the capabilities to resolve certain type of exceptions. In other cases exception resolution needs to be built into the process design for human intervention.

- Business process monitoring tool must the capability to modify live process instance data.
- Business process monitoring tool must have the capability to manually change the status and path of execution of the business process being executed.
- Process deployment tool must have capability to update activities on live processes.

I recommend companies to create **exception repositories**. Exception repositories are nothing but logging the exceptions and recovery options. In high volume processes it might be possible to generate customer FAQ's based on data in the exception repositories. Also for individuals who are new to a corporation it can act as knowledge repository to tell how a particular exception has been handled in the past. From my experience exception repositories are valuable inputs for process improvements. Process optimization phase takes direct input from process monitoring and exception management knowledge base.

Chapter 7:
Process Optimization

Chapter 7: Process Optimization 94

Continuous process optimization is the basis for existence in highly competitive markets. By continuous process improvements businesses can not only survive but they can thrive on change. Continuous process improvement is part of the culture of being a process managed enterprise. Process optimization is a journey, not a destination.

With every cycle of process optimization companies can come closer to developing innovative business processes. Companies like Dell and Wal-mart have developed innovative business processes, by continuous process optimization. These companies did not become process managed enterprises overnight with a big bang onetime process improvement project. That is the reason Compaq and Kmart have not been able to imitate the innovative process models. Process optimization is rewarding, yet painful. BPM makes it possible to determine process effectiveness and costs both before and after optimization in real-time.

> **BPMS creates real time business process metrics and provides the tools to analyze process metrics for process optimization.**

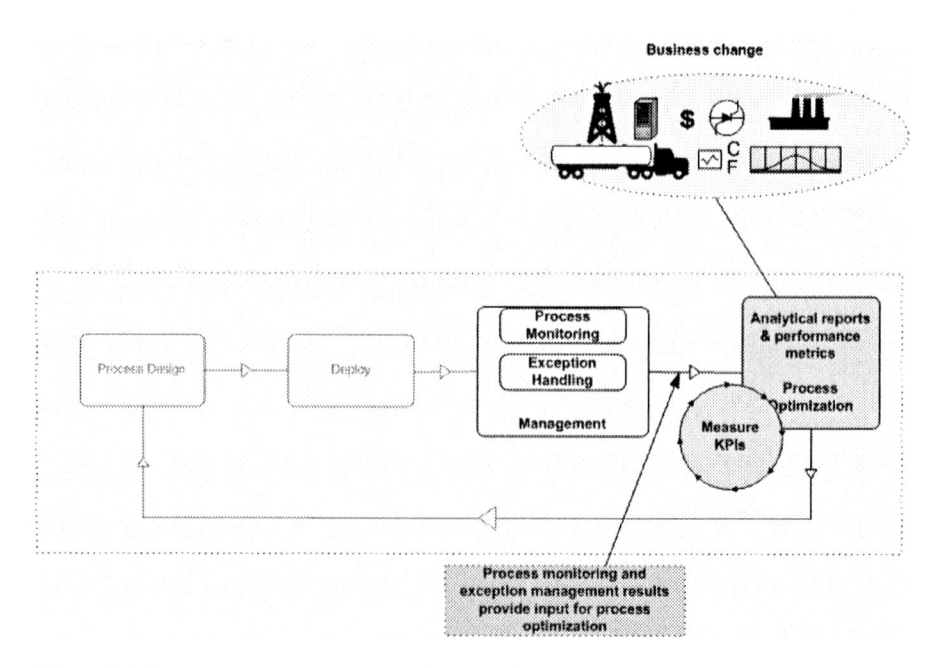

Figure 7.1 Inputs to the process optimization phase

BPMS provides essential information for understanding and improving business processes. BPMS gives business managers access to real-time process metric data. By analyzing process metrics process owners and business managers can

- Can take corrective action to realign process with strategy.
- Can take proactive action to profit from new opportunities.
- Can make timely decisions to avoid risk.

Process managed enterprises are always in feedback loop as shown in Figure 7.1. Inputs and drivers of process optimization are

- Business dynamics and change.
- Inputs that come in as a result of exception management and process monitoring.
- Continuous KPI measurement and improvement.

Are your business processes delivering the value to the customer in a cost effective and timely fashion? Is your business processes rigid, semi-automated, error prone, opaque processes which no one understands completely? Is there a disconnect between corporate strategy and grass root level implementation of the strategy? Continuous process improvement is the answer to all questions. Business process optimization is not a one time project. It is a way of working.

Business dynamics and change

Business conditions are in a state of constant change. Business processes are not just the four walls of a legal entity. Enterprise business processes need to be continuously analyzed against an array of fluid business factors like

- Customer demand and tastes
- Changes in business trends, new offering form competition
- Corporate strategy implementation
- Regulatory compliance
- Global Economy
- New technology

These factors reflect an ever changing business environment. In every industry the pace of change continues to be high and business managers need to monitor these external events. Business processes are re-tuned and re-adjusted to factor in above variables in the process optimization phase. These variables provide a set of input needed to continuously improve and optimize business processes. New business processes might be created if changes to existing processes are drastic. By

Chapter 7: Process Optimization

having a pulse on the business dynamics business managers can always align the enterprise to the strategy.

Inputs from Exception Management and Process Monitoring

The information gathered from process monitoring and exception management provides critical information about the existing running business processes and future improvements and changes to existing processes. Table 7.1 shows the difference between process monitoring and process optimization phase. These two phases of the process life cycle do not have distinct boundaries and might overlap as shown in Figure 7.2. That is the reason I explained Business Intelligence in Chapter-6. The same reports might be used with different objectives.

From my experience, some issues discovered during process monitoring are logged and passed to the BPM team for future process improvements. The exceptions the process encounters during orchestration are good candidates for process optimization. The process owners and process users must document a wish list that is generated as a result of process monitoring and exception management. This **"process wish list"** is an excellent source of input for process optimization. I suggest companies not to get bogged down with this wish list, but also think outside the box- which I will cover in process metrics and KPIs.

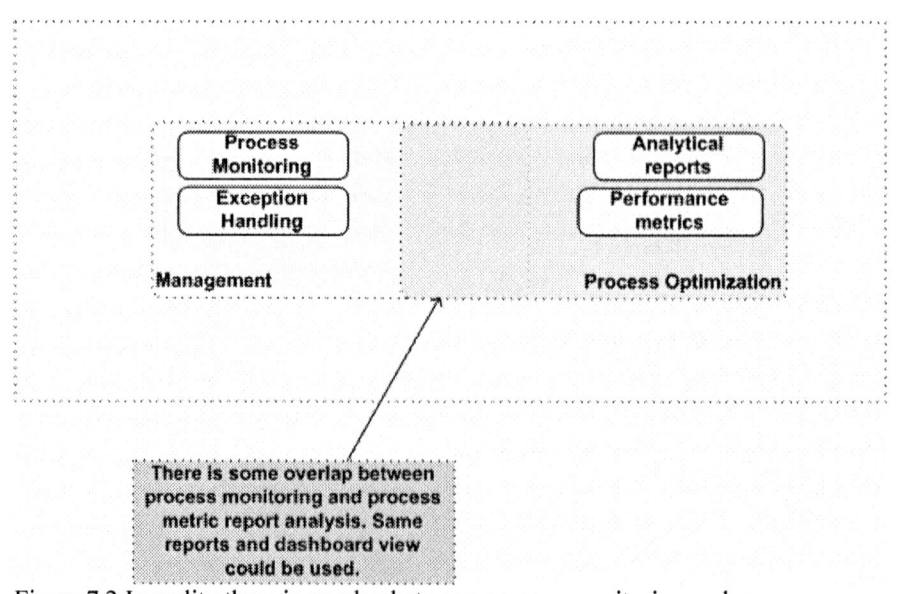

Figure 7.2 In reality there is overlap between process monitoring and process optimization.

	Time Frame	Objective
Business process monitoring	Short term (Daily, weekly)	Visibility and statistical reporting.
Performance metrics	Long term	KPI measurement.

Table 7.1 Timeframe and objectives of process management and process optimization phase.

BPM and the Balanced Scorecard

Business process management encapsulates several concepts from existing management techniques – the balance scorecard being one of them. Drs. Robert Kaplan (Harvard Business School) and David Norton created the 'balanced scorecard' system which has been used by several companies to improve performance. Kaplan and Norton describe the innovation of the balanced scorecard as follows: *"The balanced scorecard retains traditional financial measures. But financial measures tell the story of past events, an adequate story for industrial age companies for which investments in long-term capabilities and customer relationships were not critical for success. These financial measures are inadequate, however, for guiding and evaluating the journey that information age companies must make to create future value through investment in customers, suppliers, employees, processes, technology, and innovation."*

The balanced scorecard forces managers to look at the business from four important perspectives. It links performance measures by requiring firms to address four basic questions:

- How do customers see us? - Customer perspective
- What must we excel at? - Internal perspective
- Can we continue to improve and create value? - Innovation & learning perspective
- How do we look to shareholders? - Financial perspective

> **The process performance tools include, but are not limited to supporting the balanced scorecards.**

In today's digital age and in the extended enterprises process owners and managers will need to measure some or all of the above perspectives depending on which industry the enterprise operates. Traditional balance scorecard metrics are generated by aggregating data from various sources. Whereas in BPM the metrics are generated using process data generated by the BPM server which orchestrates the business process.

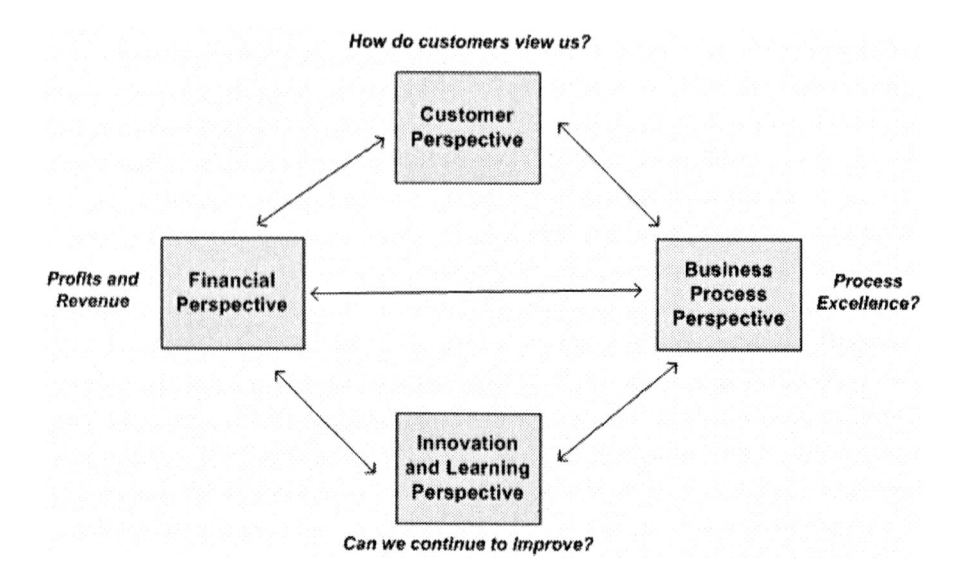

Figure 7.3 Different perspectives for process improvement

> Despite the strengths of the balance scorecard, it cannot stand alone. The reason being it can alert business managers when something is wrong but cannot provide solutions. BPMS lets business managers see the health and performance of enterprise business processes by displaying balanced scorecard metrics using dashboards and reports.

Senior executives do not need to do number crunching on the reports as process metrics can be displayed on dashboards. The idea of the balanced scorecard is to tightly link your performance measurements (financial and non-financial key performance indicators – KPIs) to your strategic planning objectives and your value proposition. Based on your industry and your business process you might have to measure one or more of the above perspectives.

> A key performance indicator (KPI) is a measurement, defined by the process owners, that reflects critical success factors for the business process. The process owners and process managers should create a dashboard view showing the actual performance of the process against the KPIs.

Customer Perspective	Business Processes Perspective	Financial Perspective	Innovation & learning perspective
Customer Satisfaction	Cycle Times	Income	Training
Retention	Quality	Profitability	Knowledge Management
Loyalty	Cost Effectiveness	Revenue	Performance
	Lag Time		
	Delay Time		

Table 7.2: Linking KPIs to perspective

There could be hundreds of process metrics that are available – but company's need to identify the KPIs that need to be improved for the business process. KPIs need to be predefined at the design phase. The key is to develop process KPIs that provide a holistic and balanced view of the business process. Process KPIs are the true "soul" of the process measurement. KPIs are key metrics used to measure the success of the business process. KPIs measure the health and pulse of the business process and ensure that all participants are moving in the right direction, working on all four cylinders towards the same goals and strategies.

Some industries have standard KPIs like the finance industry. In such cases it is easy for business owners to define the KPI and measure process performance against those KPIs. Other specialty industries or internal business processes do not have standard KPIs so process managers would have to come up with the KPIs. Even improving on one KPI say cycle time for a business process results in considerable efficiencies.

Process metric improvement

Now that we have some idea of balanced scorecard and KPIs, I will describe a few common KPIs and how BPMS enables tracking and improving these KPIs. Again KPIs are process and industry dependent and it is not possible to cover all in this book. It must be understood that since BPMS capture process data, it will be possible to measure any KPIs you want to measure.

Process efficiency metrics

Process efficiency metrics determine how effectively the designed business process is working. Process speed is crucial as customers want their products delivered on time. They want real-time information and real time status updates on their order request. Using BPM systems it is now possible to measure and calculate the time it takes to execute the business processes. BPMS allows companies to view process completion time and how it was derived. Dashboards and reporting provide process metrics which can be used as guidelines

Process completion time reports
Process completion reports show the time it took to complete process instances. Customized version of these reports can be created easily by the business people to show

- Average process completion time.
- Process completion reports based on a desired timeframe (weekly, monthly, etc).
- Process completion reports as a function of various business events.
- Process completion reports as a function of various participants.

Activity completion time reports
Activity completion reports show the time taken by individual activities to complete. Customized version of these reports can be created easily to show

- Average activity completion time
- Activity completion time based on a desired timeframe (hourly, daily, weekly, monthly, etc).
- Process completion reports as a function of various business events.
- Process completion reports as a function of various participants.

Based on the process and activity time reports companies can understand how the completion time is effected by external and internal factors, and what to do about it. Elapsed time from one activity to another can be analyzed and possible causes of delays can be identified. Delays could be due to several reasons – not enough information, uneven activity distribution among human participants, poor organization structure resulting in information jam. Business process managers are now in a position to improve process speed by taking corrective actions like automating certain manual activities, unleashing the power of parallel processing and leveraging asynchronous work calls.

Chapter 7: Process Optimization 101

> Time reports and dashboard views in BPMS, enable process owners to understand how various factors (customer type, customer, account, quarter, participant, organization structure, activity type, business event, etc.) effect process and activity time. That makes them eligible to improve the process.

Process visibility metrics

Process monitoring in the process optimization phase is about process analytics, i.e. monitoring historical process volumes to derive trends, patterns, improvements, etc. Enabling real time process visibility is an extremely powerful feature of BPMS. The mental light bulb goes on when business managers and process owners look at process data in real-time and in a graphical display. From my experience after the first version of business process is deployed there is a flood of optimization ideas which come in. That is possible because everyone can monitor the processes in real-time. The following process analytics reports are most common for further process optimization

The different reports as shown below provide guidance and visibility into business operations. The business managers can determine how their business processes are running in close to real time against an array of business factors. This will help them optimize the business processes.

Historical process reporting

Basic idea is to determine the volume and status of process instances against an array of business events, business entities or timeframe.

- Number of late processes in a desired timeframe (weekly, monthly, etc).
- Number of completed processes in a desired timeframe (weekly, monthly, etc).
- Number of failed processes in a desired timeframe (weekly, monthly, etc).
- Number of cancelled processes in a desired timeframe (weekly, monthly, etc).
- Number of processes against an internal business event. For example number of new orders during the time a new commercial was running.
- Number of processes against an external business event. For example number of cancelled flight booking when hurricane hit Florida.
- Number of process against a business entity. For example the number of new orders per customer.

Activity reports

An array of activity reports can be generated, similar to historical process reports to measure an activity or group of activities against an array of variables like time, business events, business entities, etc.

Chapter 7: Process Optimization   102

- Number of completed activities in a desired timeframe (weekly, monthly, etc).
- Number of late activities in a desired timeframe (weekly, monthly, etc).
- Number of failed activities in a desired timeframe (weekly, monthly, etc).
- Number of cancelled activities in a desired timeframe (weekly, monthly, etc).
- Number of activities against a business event. For example how many "24 hr support request" after a new software version release.
- Number of process against a business entity.

Exception reports:
Volume exception reports can be created and analyzed against an array of variables like time, business events, business entities, etc. The error reports show details like process name, activity name, error type, business condition, business event etc. Correlation can be made between various types of exception and business factors. This knowledge can be used to improve the business process.

Cost effectiveness metrics
Cost effectiveness metrics measures how cost effective a business process is. The cost of a process can be benchmarked against the value of the process. During the process design phase, cost (per hr, per day, per transaction, etc.) can be assigned to each activity. Activity Based Costing (ABC) is a very mature area of determining the cost of a business process. But without BPMS it is very difficult to calculate the true cost of a process. In BPMS – it is now possible to accurately calculate the cost of a process instance, average cost, and cost during certain timeframes, etc. Most companies are not aware of the actual cost of their processes. Dell makes money not by selling computers at a higher price – but by making its business process cost effective.

Activity cost reports:
Activity based reports provide information regarding the cost to complete a particular activity. A business process can have different kind of activities – human and automated. Cost reports and dashboard information provide cost associated with each activity against an array of factors like certain time period, business event, and business entity. The activity cost reports depicts how the cost of various activities varies with time and other business factors. This report provides insight into activity bottlenecks, most expensive activities, low value expensive activities, high value expensive activities, etc. Business process owners and managers can then improve the cost structures of the process.

Process cost reports:

Process cost reports show the cost associated with running the business processes against an array of business factors. Process managers can then measure the cost effectiveness of the business process and can improve the cost of the process. Other cost reports could be

- Average cost of business processes.
- Cost of processes with late activities.
- Cost of processes with cancelled activities.
- Cost of processes against different partners and suppliers.

Resource planning metrics

Tracking and managing the knowledge based workforce has become a challenging issue faced by companies today. In today's knowledge based economy human resources are a companies greatest assets and the largest cost center. Effectively utilizing human resources is essential to efficient and cost effective business processes. Using business performance tools business process managers can get a view of the current resource utilization from a process context. The objective of resource planning is

> **In today's knowledge based economy, as the workforce becomes more specialized and diverse in their skills and activities, more mobile, and more geographically dispersed, traditional resource planning management methods are no longer adequate. Getting the most out of your workforce is critical to the success of any operation since due to lack of visibility, inefficiencies can creep in and lead to poor customer service and loss of revenue.**

- Measure and analyze employee activities to drive organizational change, control costs, and increase employee productivity.
- Monitor how employees are actually spending their time across all processes and activities.
- Empower employees to track and manage their own activities, ensuring greater awareness of their productivity.
- Stay focused on achieving day-to-day operational goals and long term optimization.
- Incorporate adherence, skills, and proficiency information into future schedules
- Understand the skills and level of proficiency of human process participants across processes.

Human activity reports:

Human participant activity reports provide details on the number, status and time of activities by individual users. Using business performance tools of the BPMS system, department managers and process managers can slice and dice, drill down information about human activities for effective resource planning. These reports are also useful for evaluating performance of individual users. The following reports can be generated in this category

- Number of activities performed by each employee
- Time per activity per employee.

Workload activity reports:

These reports provide information on what work is pending for one or many human participants across one or many processes. Workload reports can be generated to show active, in progress, late and urgent processes and activities. Some of the following information can be viewed by reports and through dashboards

- Human participants based on activities.
- Time taken to complete activities.
- Pending work items by human participants.

Process Optimization

Process metrics reports provide actionable data in real-time relevant to the process. Real-time statistics about productivity, cost, completion times, bottlenecks, process anomalies, workloads, etc empower business managers to always stay on top of issues and manage the business process. Process measurement provides critical information regarding customer needs, partner and supplier performance, organization flexibility and helps in strategic decision making. Bottom line is with BPMS the processes are orchestrated electronically and every event and metric is captured and made available. The following process improvements can be done

- Identify possible bottlenecks in the process. The bottlenecks could be due to several reasons like human work load, process design and process flow problems.
- Certain suppliers are not executing as per recommendations. They need to be replaced other service providers.
- Process owner could discover sequential execution of unrelated activities is causing a slowdown in process completion time. Process design could be changed to make unrelated activity steps to execute in parallel.

- Bottlenecks could be due to information getting jammed at one activity because it requires the feedback/approval of one important person. The process owner could then modify the process accordingly.
- Bottlenecks could be time related, say during "back to school" timeframe processes start getting delayed due to poor resource allocation.
- Resource allocation can be changed to improve process time.

Once key areas which need to be improved are clear, the improvements are factored in the business process design. The feedback loop is always in action.

Chapter 8:

EAI (Enterprise Application Integration) and BPM

Chapter 8: EAI (Enterprise Application Integration) and BPM 107

Traditionally enterprises grew their systems and IT infrastructure out of necessity as the business grew and connected the various IT applications with minimum planning. Stove pipe applications were pretty much the norm. These applications were connected in a point to point manner with no clear indication of process flow. But as the pace of business change accelerated connecting systems became a major roadblock. Back in Chapter 2 I wrote about the disadvantages of stove pipe applications and point to point integrations. As shown in figure 8.2, in large organizations point-to-point integration has created spaghetti architecture. I have consulted with clients who have not been able to retire legacy systems because of the spaghetti architecture problem. Stove pipe applications support business processes in a fragmented and discontinuous manner.

Stove Pipe Applications: Each application is heterogenous, has it's own datastore and produces documents that are shared between business users to run the busines process

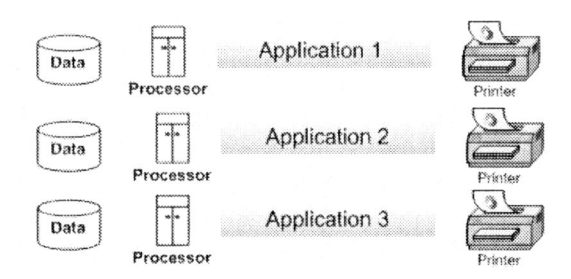

Figure 8.1: Stove pipe application architecture.

Point to Point Integration: Integration is rigid, expensive and provides no indication of business process

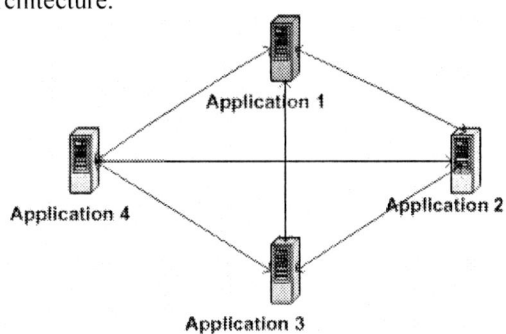

Figure 8.2: Point to point integration

Then in the late 1990's EAI (Enterprise Application Integration) emerged as a planned way to integrate the array of IT applications. EAI became a top priority in many enterprises. With the internet and the creation of the extended enterprise – B2B (Business to Business) integration became possible.

EAI (Enterprises Application Integration):

Sophisticated EAI suites like BizTalk, Tibco etc came to the market. However despite increasingly sophisticated EAI applications, enterprise application integration remains difficult and does not meet business objectives. Enterprise integration has to deal with multiple heterogeneous applications running in different locations. Figure 8.3 shows a typical EAI architecture within an enterprise. Using publish and subscribe technology the EAI system senses any data change in the source system and updates the destination systems. Also scheduled batch jobs are used to transfer data. Figure 8.4 shows a B2B (Business-to-business) integration architecture. However EAI failed to deliver its promise and some of the below business issues still persisted

- Low visibility of business operational performance.
- EAI is data centric and not process centric.
- EAI could not keep up with business process change.
- EAI does not address the business process.
- EAI solutions are technically very complex, need specialized skills and are very expensive to maintain.

Basically EAI addressed the problem of connecting applications and transferring data. EAI is not going away. With BPM the focal point is going to be the process and not data. BPM will leverage the EAI investment made by corporations.

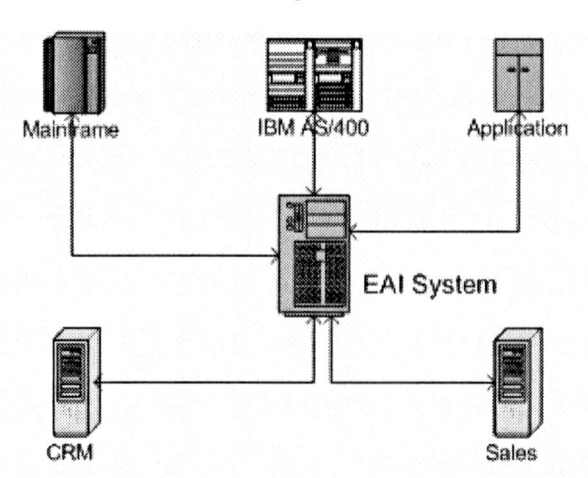

Enterprise Application Integration within the enterprsie
Figure 8.3 Depicts internal EAI architecture connecting heterogeneous systems.

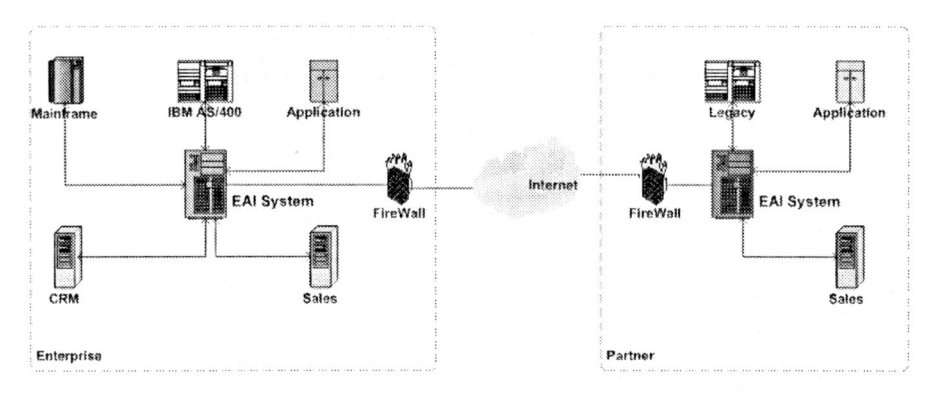

B2B (Business to Business) integration with partners and suppliers.

Figure 8.4 Depicts B2B integration architecture connecting heterogeneous systems across the enterprise.

EAI in BPM

Traditionally the focus of EAI was data and data change. EAI connected applications around data. Data is the first class citizen in EAI. In EAI anytime data changed in the source it was automatically propagated to other connected applications. BPM leverages the application connectivity aspect of EAI but changes the focus to process. The business process as it executes uses EAI technology to connect to disparate systems and propagate data. Here is the table from Chapter 3 regarding the difference between standalone EAI and EAI in BPM

Capability	EAI	EAI in BPM
Data Flow	Automated Application to Application Data synchronization.	Business Process could involve both Application to Application data flow and human interaction to enter/review/update data in the flow.
Duration	Simple and short running flows of Data	Business Processes could be very complex and long running.

> EAI is a subset of what a BPM system offers. BPM products must have built in application integration capabilities so that existing applications can be leveraged. The most optimized way to integrate with existing applications is using Web services. Nevertheless BPM products need to provide integration capabilities built in and in partnership with third-party EAI products.

For complex business processes where multiple heterogeneous applications are process participants the ability to integrate and orchestrate the applications by the BPM system is critical.

Web Services

The internet has created universal connectivity. The extended enterprise uses the internet to connect to its customers, partners and suppliers. Web services have forever changed the way we think about enterprise application and architectures. Web services allow language, platform and location independent connectivity using open universal standards like HTTP. Web services allow heterogeneous applications to talk to each other without expensive EAI software. Web services enable existing application to be exposed as services. SOA (Service Oriented Architecture) is a collection of loosely coupled web services that communicate with one another. SOA where business logic and business processes are exposed to other processes and software through standard services.

> BPM and web services compliment each other. Web services can be used by the BPM system as a way to integrate other IT applications in a process centric way.
>
> Secondly, the BPM system itself can be exposed as a Web service and other applications and systems can initiate a process through the BPM system.

Chapter 8: EAI (Enterprise Application Integration) and BPM   111

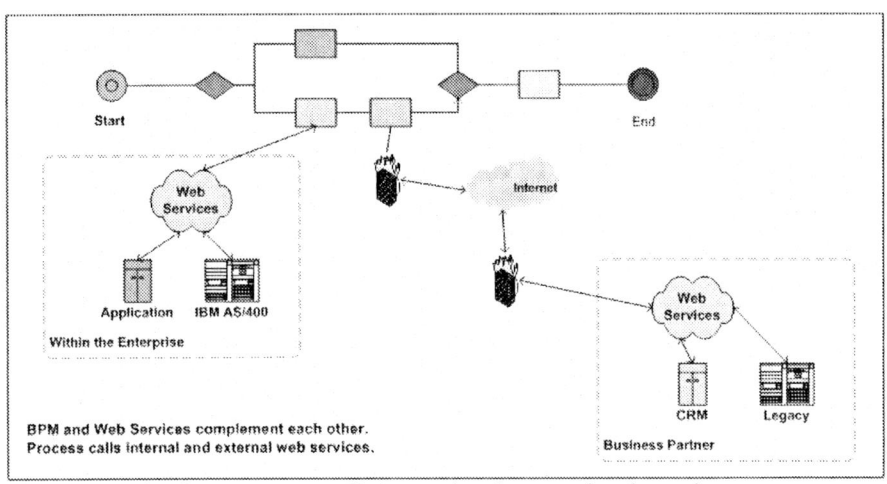

Figure 8.5 A business process calls internal and external web services.

Figure 8.5 shows a business process invoking internal and external web services. External web services are invoked over the internet and can be used to collaborate with external partners and suppliers. Trust me BPM and web services are a powerful combination and when used together the combination creates mind blowing flexibility, loosely coupled integration and excellent opportunities for collaboration and business process outsourcing.

BPM and Web Services – a powerful combination

BPM and web services will dominate the next few years. This combination creates a rich matrix of process possibilities which if I where to describe in a few words I would say "process everywhere". BPM and web services facilitate a new breed of process automation and innovation in the extended enterprise. The dynamic combination of BPM and web services can truly create the following process possibilities

- Pay per user process architecture.
- Process Plug and Play architecture.
- Process embedded within processes.
- Universal business processes.
- Specialized business processes.
- Process services.
- Process collaboration.

Web services are becoming a major catalyst for the evolution of BPMS. The ability to dynamically invoke web services in a synchronous or asynchronous manner from within the business process during process orchestration is enabling the creation of new industries and process innovation. In fact the cost and complexity of BPM solutions can be reduced through the use of web services. There are certain challenges that need to be kept in mind for complex business processes like

- Coordination of asynchronous communication between the business process and the web services.
- Long running transactions during process orchestration.
- Correlate message exchanges between the business process and web services.

EBay is the worlds biggest online auction site. Thanks to internet, rock solid auction process and web services eBay has converted itself to the biggest flea market of the planet. Nearly 40% of the product listing are generated by web services. By making eBay accessible through web services, business are empowered to leverage eBay's trading platform and auction process. Businesses are embedding eBay's auction web service into their own business processes like traditional sales, delivery and procurement processes. This new plug and play process and web service architecture is generating efficiencies and profits for forward looking companies.

> **Processes managed enterprises are creating new industries by making available internal business functions as pay-per-use processes and creating new industries.**

Figure 8.6 shows how eBay's auction process is exposed as a web service. Vendor's delivery and procurement processes call eBay's auction process over the internet using web services. This real time example shows process "plug and play" and "pay-per-use" process architecture. Vendors now can embed eBay's super efficient auction process into their own business processes and give eBay a service fee. EBay's auction process communicates back to the vendor's process and the vendor has complete visibility of product auction and complete control. BPM and SOA together will create virtual enterprises like eBay. The point is organizations do not have to create everything on their own. They can create innovative processes that contain sub processes created and managed by others.

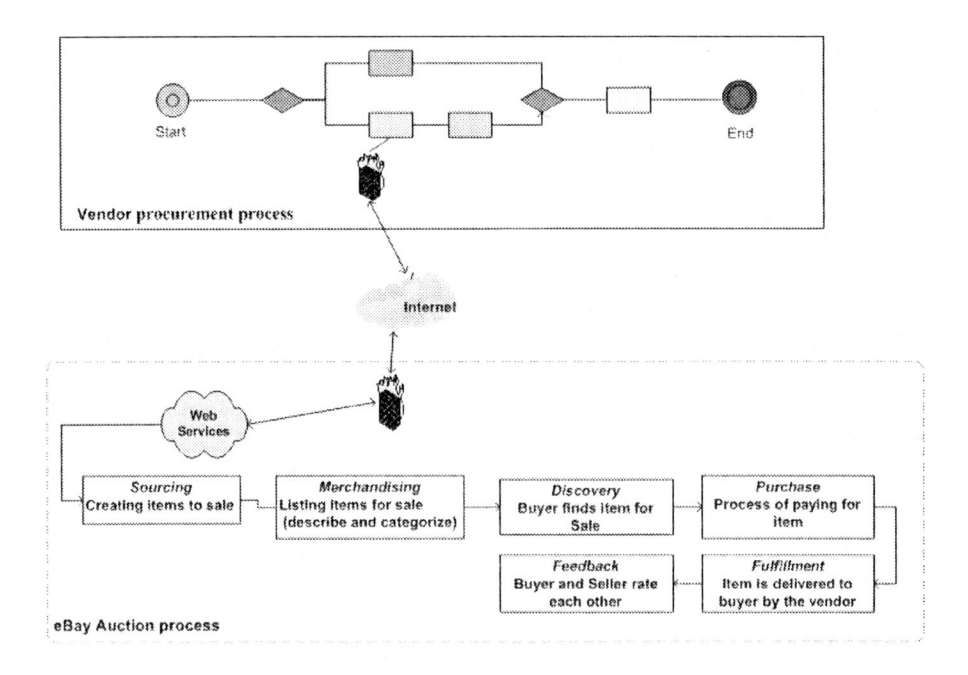

Chapter 8: EAI (Enterprise Application Integration) and BPM   *113*

Process Plug and Play Architecture.
Pay-per-use Process Architecture.
Figure 8.6 EBay's auction process exposed as a Web service

Using BPM and Web services in tandem may allow the extended enterprise to extract far more value out of internal applications and leverage external business processes like we have seen in Figure 8.6. As BPM moves towards mainstream adoption, we are going to see an evolution of business processes and services. Web services will quietly merge behind the scenes with BPMS, and we are likely to see processes every where. Universal services like currency rate services will be offered and maintained by some company. All other business processes will just consume the currency rate service.

Business process and web services combination is already changing the business landscape. Business processes will invoke web services, but it might be the end consumer who is interested in the results of the web service. Dell might use web service from UPS in its direct process model to offer its customers package tracking information. Dell customers can go to UPS web site to track their packages once Dell gives them a confirmation number. Such kind of process handoff and process specialization is going to be the norm going forward.

Chapter 9:

Collaboration

In the previous chapter I wrote about B2B interaction and collaboration. In this chapter I will focus on people to people collaboration. Complex business processes are long running orchestrations and might require human intervention and control. In today's service economy worker knowledge is a tremendous asset. Knowledge workers need to collaborate to service their multinational customers, keep value chains running, produce innovative products and services, comply with regulators, etc. People in the extended enterprise need to collaborate around the business processes across internal, external and geographic boundaries with colleagues, partners and suppliers. Collaboration is fundamental to ensure enterprises operations are efficient, effective and unified. For example to offer customized solutions for it's customers the sales, service, product and marketing departments personal might need to collaborate.

Collaboration around the business process falls within one of the two categories: synchronous or asynchronous. Synchronous collaboration involves people to be present physically or virtually while collaborating like instant messaging. Asynchronous collaboration like discussion threads allows participants to post their ideas on a digital board. Collaboration supports business processes by facilitating interactions between process participants, discussions to develop around key issues and knowledge to be collected for reuse. From a business process management perspective human participants will need to collaborate around the running process to get the job done and handle any exceptions that may arise. Collaborating around the business process has tremendous benefits

- Greater knowledge sharing resulting in accelerated process completion.
- Consistent and unified service to customers and clients.
- Reduced human error and faster decision making by leveraging expert opinion and knowledge sharing
- Cost effective and efficient process completion through shared process intelligence and resources
- Efficient processes due to greater decentralization of certain decision rights and increased delegation. This is possible as process managers have the visibility of the real time process and can collaborate around the process with their subordinates.
- Team oriented process completion resulting in process improvements and optimization ideas.

Typical collaboration techniques are sharing and managing of documents, discussion list, discussion groups, instant messaging and emails. Whereas email has become the de facto communication standard we are more interested in collaboration around the live business process. What we are interested in is the persistence and archival of the resulting process and knowledge intelligence for future re-use.

> Real world complex business processes have human steps where "negotiation" among various process users and stakeholders eventually leads to the execution against deliverables. Such negotiation human activities are hard to automate and require high degree of collaboration.

BPM and documents

Most digitized business processes have documents as inputs and/or outputs of the business processes. Documents are not shared just for the sake of sharing. People want feedback and edits. Some documents could be read only like pictures, PDF, etc. For compliance type business processes documents are the output of the business processes. These documents are submitted to regulatory authorities for compliance. Some business processes produce Service Level Agreements as output documents. Managing these documents around the business processes is a critical element of BPMS.

> BPM products must provide a mechanism to route the correct documents to various process participants. Advanced BPM products must have the capability to integrate with existing DMS (Document Management Systems).

Providing the capability to associate documents with the process or activity is critical. Documents should be routed in a reliable and secure manner. Passing and receiving documents across processes is a key requirement of BPMS. Advanced BPM products allow integration with existing DMS like documentum, sharepoint etc. BPMS is a technology platform which places the process at the center- with all services including document sharing revolving around the business process.

Discussion Threads

In today's knowledge based economy where complexities are rising people participants of live processes need various mechanisms to collaborate. People participants located across the globe, across the enterprise need various robust and secure tools to collaborate. BPM system must provide the tools and means to collaborate. BPM systems must link and persist the unstructured collaboration that happened during the life of a process instance. Many BPM products do not provide such extensive means of collaboration at the present, but as the BPM landscape evolves I think people collaboration will become integral part of BPM systems. BPM vendors do not have to build from scratch collaboration technology into their products. Collaboration technology can be provided by bundling existing collaboration products with in the BPM product. The following types of discussion threads in a BPM system will immensely enhance team working and sharing of ideas

Discussion groups around the live process instance

Process participants must interact with other process participants and non participants scattered across geographical areas. Discussion threads allow process level discussion among participants and non participants to share process business knowledge and information. Non participants could be invited to participate in the discussion group for their expertise. For example participants in the "advertising campaign process" could invite the legal department representative in the discussion group for advice on legal issues although the legal department is not involved in the advertisement creation process. Process discussion threads allow

- Establish discussion groups on any topic to generate feedback or new ideas from your group.
- Post a suggestion or problem solution in a message thread.
- Brainstorm ideas for procedures and proposals.
- Clarify complex issues.
- Search a particular discussion thread by keyword or title.
- Subscribe to messages in a particular group.

Figure 9.1 depicts human participants collaborating using discussion groups and real time technologies like instant messaging around the process instance. In the extended enterprise as the workforce becomes more mobile and more geographically dispersed synchronous and asynchronous collaboration among process participants is critical.

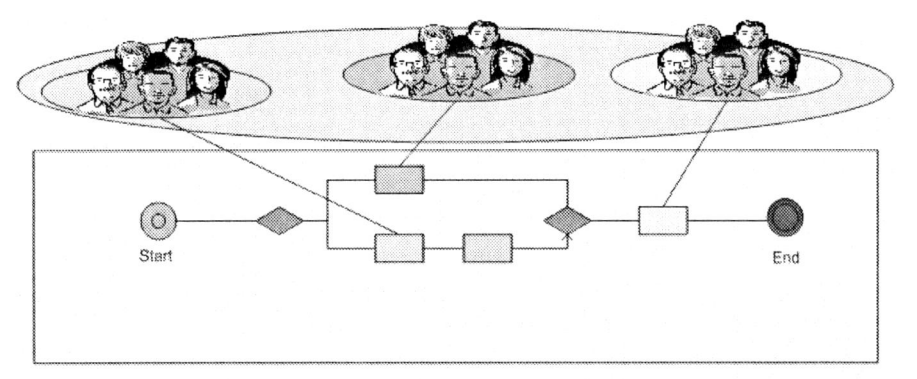

Collaboration around the process instance by human process participants.

Figure 9.1 Collaboration in a live process using various collaborative technologies.

Discussion groups around activity instance

In certain cases human recipients of a particular activity might need to collaborate in relation to that activity and they would like to keep their collaboration private. For example a particular work item might go to a *Task Queue*. The group of process users responsible for work items in that task queue might be located in different offices across the globe. This group of process users might need to collaborate to share knowledge. The BPMS must have the capability to create activity level collaborations as shown in Fig 9.2. The following features will be helpful in collaborative environments

- Ability to mark process level discussion threads private so that only process participants can view/search them.
- Allow process participants to create discussion groups at an activity level.
- Allow process participants to make discussion groups at an activity level private.
- Allow process participants to invite non participants for collaboration.

> Process level and activity level discussion threads persist with the process and can be archived for business intelligence and future process reference.

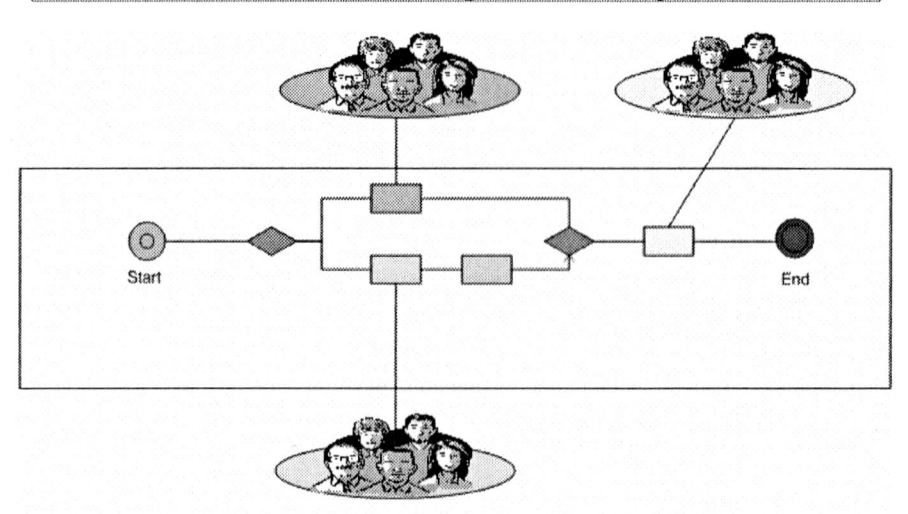

Collaboration using discussion groups around activities.
Figure 9.2 Collaboration among activity recipients.

Innovative process managed enterprises like Dell are using this simple technologies to cut big costs in their customer support process. Figure 9.3 depicts how Dell uses discussion boards as a mechanism for customers to serve other customers in need of assistance. Basically what Dell is doing is making available

Chapter 9: Collaboration *119*

intelligence that was generated as a result of previous customer support instances and also allowing customers to post messages if they have issues. A Dell moderator monitors these discussion threads and sometimes provides much needed answers to questions. Moderator also monitors these customer contributions for accuracy. Dell is profiting from this simple technology in several ways

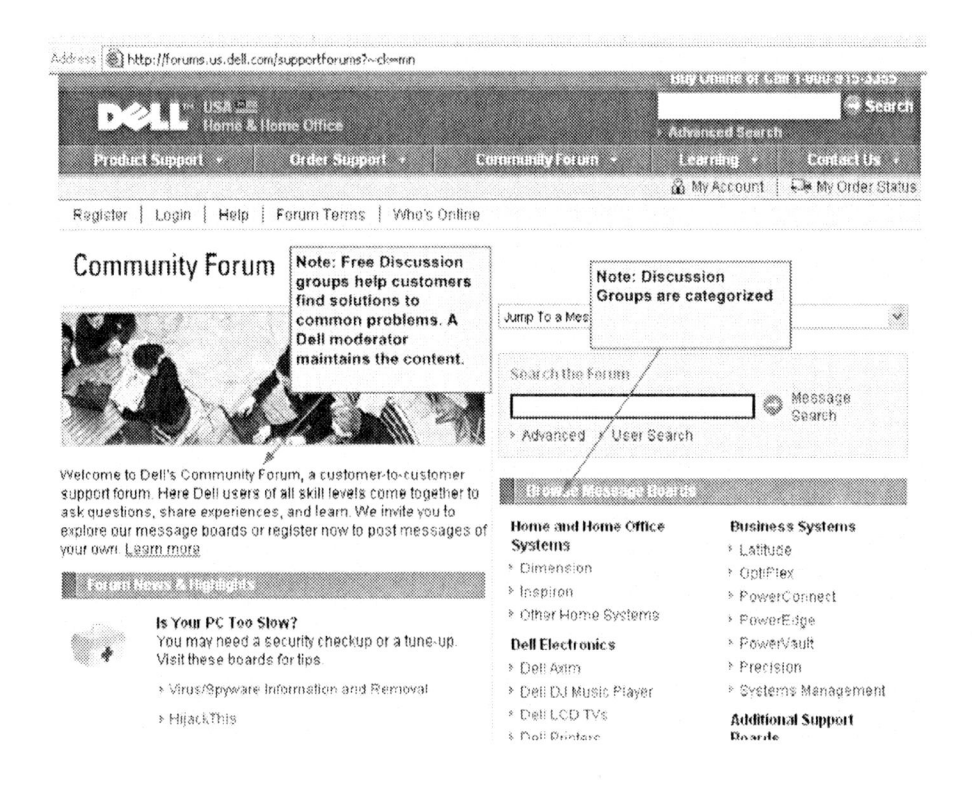

Figure 9.3 Dell Customer support uses customer-to-customer support using discussion groups.

- Cost cutting because customers find answers to most rudimentary issues on these discussion threads and do not need to contact Dell customer specialist.
- Keeps customers happy because many technology savvy customers do not want to go through the rigmarole of the customer support process.
- Customers who do not have a customer support agreement or do not want to pay for support can get their answers right there.

- Customers can subscribe to the message boards and get notified via email if there is a new message or thread.
- Customers are empowered and feel in control.

As seen above collaborative intelligence in the form of discussion boards and threads can be unleashed to drive efficiencies, reduce cost and increase customer satisfaction. In customer oriented business processes and internal business processes process owners or someone can be responsible for categorizing and arranging this knowledgebase which was generated when the process was executing in a public area or the portal.

EBay has gone miles ahead with a robust member-to-member feedback system for their auction process. EBay has created a feedback forum. The feedback forum is the place to learn about your trading partners, view their reputations, and express your opinions by leaving feedback on your transactions. Such member-to-member comments help the millions of buyers and sellers in the community build trust and share their trading experiences with others. Based on member feedback eBay calculates a Positive feedback score which members use to gauge the trust worthiness of sellers and buyers. Members feel empowered as they have the power to rate other members. The system has worked and eBay has become the world's largest flea market.

Real-Time Collaboration

Integrating real-time collaboration like instant messaging in BPMS allows process participants to collaborate in real time, quickly discuss ideas and issues, resolve issues and make joint decisions with regard to the business process. Real-time collaboration among process participants increases the process efficiency by removing the time roadblock waiting for email or phone calls. Instant messages can be saved with the process instance as a discussion thread for future reference. **Presence aware routing** in BPM products routes work items based on who is currently online, thereby increasing the efficiency of the process.

BPM specific collaboration

The BPM server orchestrates the business process as per the process design. At the same time the BPMS must provide capabilities to the human process participants to control, delay and reverse the process orchestration for collaboration purposes. The following examples make the above point clear

Returning work items
The BPMS must have built in capabilities for human participants to return the work activity to the previous participant when he/she decides that the information is incomplete. The previous participant then completes the information and resends the work activity forward. In complex processes like R&D a work activity can go back and forward many times as per the discretion of the process participants.

Sending work item for reference
Human participants might have a need to send a work item they have received to a non process participant for feedback or advice. For examples as mentioned before a participant might want to send the work item to a member of the Legal team to review the wording of a campaign. The BPMS must provide such kind of services.

Increasingly as BPM products mature collaboration with the process at the center will become an integral part of BPMS. In the future some BPM products will have collaboration capabilities out of the box and/or will provide integration with industry collaboration suites.

Chapter 10:

BPM Servers

Chapter 10: BPM servers *123*

There is too much of confusion out there regarding BPM, BPM servers and BPMS. Software vendors, VARs (Value added retailers), strategic consulting firms, research analysts have been providing different meanings for the above terms. I want to get the terms straight before I dive into details of BPM servers. BPMS (Business Process Management Systems) are complete technologies which enable the business to manage the business process. A BPM server provides certain services like EAI, orchestration, workflow, etc which are needed for process orchestration and process management. Below is the concept BPMS diagram from chapter 3. As we see BPM engine/server is the core of BPMS.

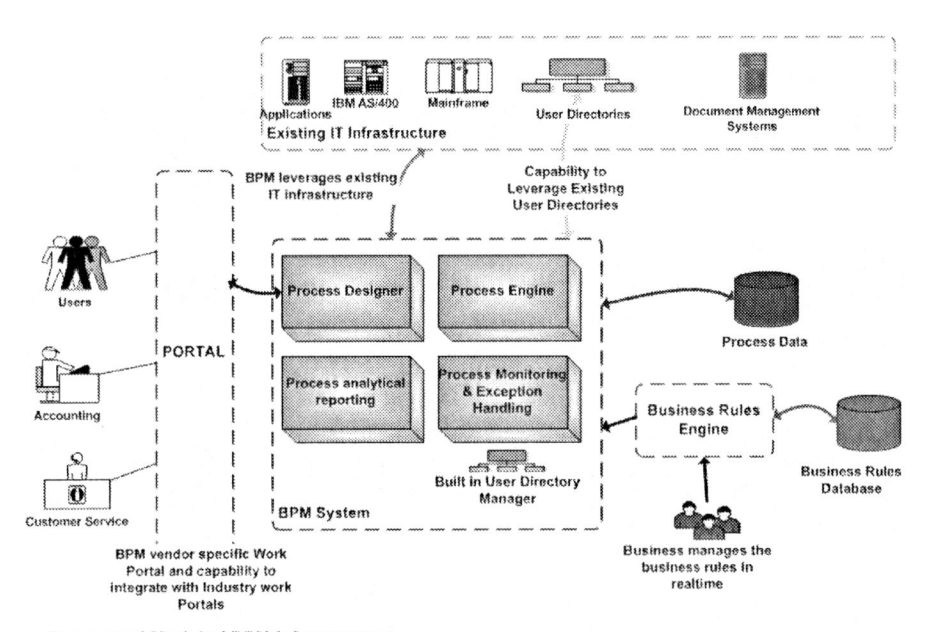

Conceptual Model of BPM Infrastructure
Figure 10.1 BPM concept model

The goal of BPM server (the process engine) is straightforward yet very complex- to provide all services needed for process orchestration, enterprise integration and collaboration. We do know that complex business process involve multiple participants (human and system), cut across the departments of an organization, involve partners and suppliers and are long running. Now think about orchestrating this business process start to end in a reliable and secure manner. To achieve this, a BPM server combines several different preexisting technologies (EAI, workflow, collaboration, etc as mentioned in chapter 3) around the business process. BPM server relies on different kinds of technologies and is distributed in nature. The federation of all these technologies targeting business change is very powerful and when done right creates a process managed enterprise. BPM server offers an array of services for robust process orchestration as described below

- Orchestration Services
- State Management services
- Transaction Services
- Management services
- Business rules services
- Enterprise Application Integration services

Orchestration Services

When a business process has been deployed, it is the BPM server which orchestrates the business process. The business process is designed by the BPM team by graphical process design tools but these diagrams are not executable. When a business process design is deployed, the BPM server translates this diagram into a format which the BPM server can orchestrate. Today different BPM products have their own proprietary mechanism to orchestrate web processes. Trying to run a BPD on different BPM servers is not possible unless the BPM vendor supports BPMN and BPEL (Business Process Execution Language). BPEL is a standard being developed by a group of vendors. In the future when all BPM vendors start supporting the BPEL standard (which itself is still evolving) it will be possible to orchestrate a business process design using any BPM product.

> **At that point in time we will experience a process revolution where process plug and play, process collaboration, process reuse, process patent, process sharing, process leveraging, process providers, process monitoring providers, process banks and process consistency will be norm. Process will be similar to what data is today.**

Business process orchestration offers unprecedented and revolutionary advances over process automation. Business process orchestration offers the ability to manage extremely complex business processes with efficiency, flexibility and agility than other wise possible. Do not confuse between web services and process orchestration. Web services can act as loosely coupled activities (if designed right) and the BPM server (process engine) orchestrates the process. The BPM server stores the results of process orchestration in a production process database (BPM Database) with all process metric details like – process status, activity status, start time, end time, delay time, notifications, etc which is collectively called process data. We have already discussed the value of process data as indicated in process monitoring, optimization and exception handling phase.

- The process engine controls the state and execution of each live process instance.
- When a process is instantiated work items are generated based on the process flow, participants and business rules. The work items are then distributed to the appropriate participants by the BPM server.
- The BPM server manages the sequencing and distribution of activities, time constraints on activities and the whole process and the state of activities. Yet the performance of work is left to human workers and IT applications. The process engine determines the sequencing of steps at runtime based on available data.
- The BPM oversees the execution of "composite" business processes that span applications, departments, and organizations.
- The BPM server alerts the users and applications when an action is required, whether to manually resolve an exception, work items which are due or getting late. The BPM server automatically records every transaction to make it's process data available for process monitoring.
- The BPM server can **pause** process execution and listen for internal or external events specified in the process design. Once matched the process orchestration would continue as per process design.

State Management

Most complex business processes are long running transactions. Process orchestration could last for days, weeks or longer. This is the biggest difference between BPM servers and application servers or Web services. Request to web services/application servers generates a reply within a few seconds or minutes.

The BPM server manages the state of all the running process instances. The **state** of a live process at any point in time refers to where the flow is at the current time, path of execution completed, events raised so far and all details of activities that have happened so far. Given the need to track process instances over long periods of time, the BPM server stores the state in a process database. It is not possible to keep in-memory state information for extended periods of time.

While a process instance is waiting for something, say user to complete approval or an event to occur in an external system, the BPM server no longer stores the process instance in BPM server memory. The BPM server saves the state out to a process database. The process instance is said to be *quiesced* (in silent state). Periodically (based on frequency defined by BPM teams which could be 30 or 5 minutes) or when an event occurs, the BPM server retrieves the process instance from the process database and orchestrates the process instance.

In the event of a BPM server shut down, the BPM server resumes activity where it left. The snapshot of all the live processes and completed processes is not affected when BPM server is brought back online as all process state is stored in the process database. Business users will get the same snapshot of all their work items when BPM server is brought online. State management is another difference between BPM servers and application servers.

Transaction Services

Process is the enterprise. The enterprise stops if the business process stops. The enterprise cannot effectively deliver value added services to the customer if the process orchestration is compromised in any shape or fashion. BPM server must provide highly reliable transaction services for process orchestration.

As discussed above, the BPM server saves the process instance to a process database for state management. This requires a transaction to be committed. The BPM server will orchestrate the process up to a point where the instance state has to quiesce.

The BPM server must merely perform the process data update (to quiesce the process instance) and return control to the caller. The BPM server must provide an exception handling capability so that exceptions can be trapped. That way it is the responsibility of the activity instance to decide to abort or commit.

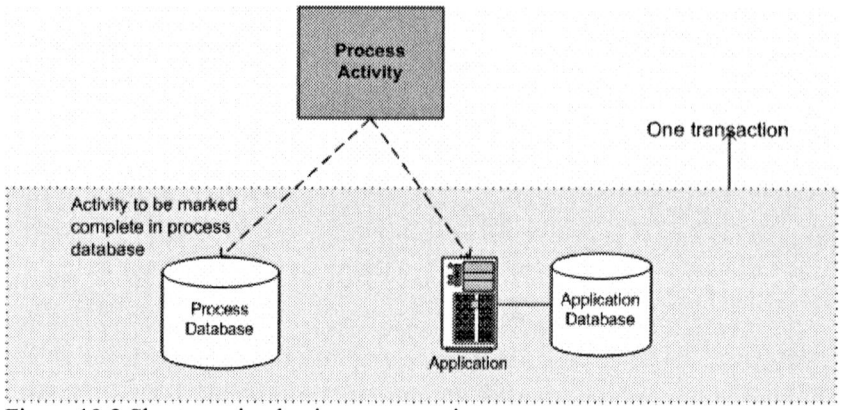

Figure 10.2 Short running business transaction.

For example in figure 10.2 the process activity updates application data upon completion. It is desired that application data update and process data update (which marks the activity complete) to be one transaction for the following reasons
- If application data could not be updated successfully, activity should not be marked complete. This gives the user another chance to enter data and update. Programmatically process data update would have to be rolled back.

Chapter 10: BPM servers *127*

- If application data update was successful, but process data could not be updated successfully. In this case BPM server should communicate to activity through an exception handler and application data update needs to be rolled back.

Above example indicates the transaction services through exception handler which BPM servers must provide. In this case atomic transaction is possible. Figure 10.3 shows a process where it is desired updates to database1 and database2 be treated as a one transaction. Since business processes are long running, the time between activity1 and activity2 could be long. BPM server would have to lock distributed data for long periods of time, thereby having a negative effect on performance and stability, and deadlock problems. I think BPM servers do not provide such kind of transaction support. This type of issue is best handled at design time by **compensating activities.** Compensating activities run if a failure occurs and undo undesired updates.

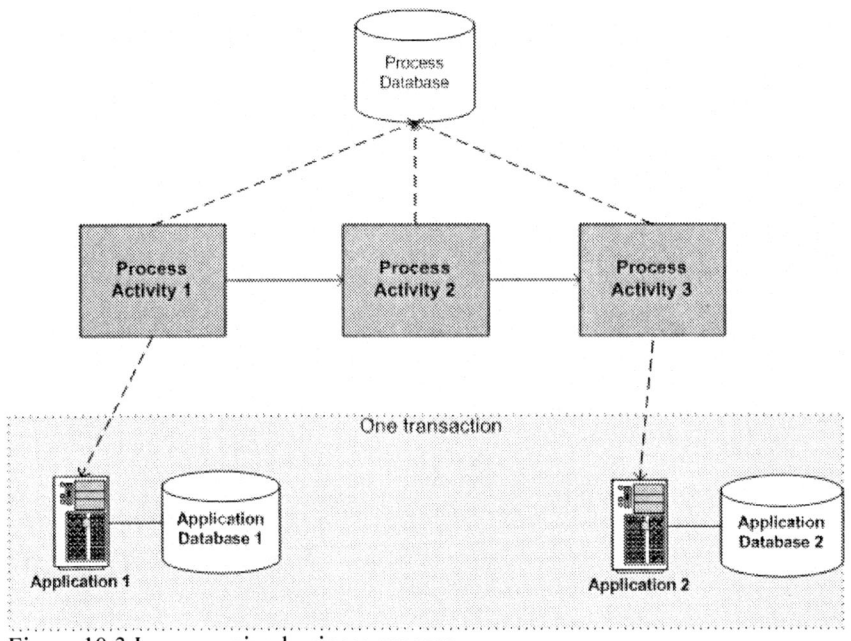

Figure 10.3 Long running business process

Management services

Business process monitoring allows business managers and owners to monitor and manage running processes. That is the business aspect of process monitoring performed by process performance tools. Here I am writing about BPM server monitoring. BPM server management is the technical aspect of management services where IT administrators can view and mange the health of the BPM server. Since BPM server implementation is vendor specific, different BPM products have different ways to manage the BPM server. Apart from monitoring the health of BPM server other management activities as mentioned below are done using the **BPM server administrator tool.**

- Deploy business processes to different process databases.
- Remove business processes from BPM server.
- Examine BPM server logs/ statistics to examine health of BPM server.
- Examine BPM server Load and BPM server utilization.
- Delete process instances.
- Transfer business process across various process databases.
- Other technical administration based on BPM server offering.

Business rules services

If you are operating in a non volatile industry where business rules are fairly static it might make sense to embed the relatively few business rules into your business process design. But if you are operating in a highly volatile industry like telecom or retail it might more sense to take the business rules out of the business process into a business rules engine (BRE). BPM servers must allow integration to standard third party BRE and the BPM product must come with a built in BRE. That way the business process being orchestrated can query the BRE on a need basis.

Enterprise Application Integration services

BPM leverages not replaces existing IT systems. BPM servers need to provide EAI services to communicate to a wide array of existing applications. BPM servers will provide EAI services through a multi channel approach
- Enables connectivity to a wide range of enterprise applications through a standards-based interface.
- Provide built in integration capabilities like script API to invoke existing IT applications.
- Native connections to common communication products.
- Support and integrate existing industry wide EAI products.
- Support Web services.

In the future the dominant way to integrate existing IT applications in the business process will be via Web services. Refer chapter 9 where I wrote about how cutting edge firms are already using a SOA (Service Oriented Architecture) approach for process management.

Chapter 11:

Process Excellence

Process managed companies are pushing the strategic envelop and expanding in times of recession, creating unique and new industries out of their internal business processes, positioning themselves ahead of the competition, writing new rules on customer satisfaction and enabling the virtual enterprise. The process managed enterprise is crushing competition (Wal-mart vs. Kmart and Dell vs. Compaq) and forcing suppliers, partners and industries to increase efficiencies.

The process managed enterprise can embed in your organization and become an integral part of your organization. The process managed enterprise is creating a flat world and sends and executes work wherever it's executed effectively and efficiently. The process managed enterprise is loved by the customers as it empowers the customers to drive and dictate. Process managed enterprise is a strategy execution machine and can make turns and u-turns on a dime. You will lag and struggle for survival if the process managed enterprise is your competitor but you will reap benefits if the process managed enterprise is your partner. I will write about two process managed companies.

Process managed enterprises orchestrate value and are constantly evolving chains.

Agility ↑   *Visibility* ↑   *Cost* ↓   *Efficiency* ↑   *Risk* ↓

Figure 11.1: Attributes in a process managed enterprise.

Process managed enterprises orchestrate activities, web services and other processes at many levels. Wal-mart orchestrates its suppliers and their suppliers in it's supply chain process. Wal-mart has shifted responsibility to re-stock shelves to its suppliers and their suppliers by refining and optimizing its supply chain process.

"Our supply chain could not deliver the services Wal-Mart expected," says Levi's CIO David Bergen, who spent time at Wal-Mart's Bentonville, Ark., headquarters during "exploratory meetings" before a deal was signed. Being a supplier to Wal-Mart demands a certain level of performance—and cost control. Wal-Mart drives you to work with your supply chain to put the same requirements on your suppliers that Wal-Mart puts on you. If you can't make your supply chain work, you won't benefit from being a supplier. Period.

Process managed companies can innovate products and services faster than their traditional counterparts. They do not have to build assets ground up. Process managed enterprises leverage assets and processes of other companies creating a agile value chain. Due to continuous process improvement, visibility and benchmarking they know the industry better than suppliers and partners. The benchmark supplier processes and provide continuous feedback. Faltering suppliers and partners are quickly removed from process orchestration and new ones are added. That ability to be loosely tied in a process managed enterprise reduces the orchestrator's risk. I write about two companies which are benefiting from being process managed.

Many industries are being hit by the commoditization wave, where prices start to decline and making profits becomes difficult. The whole industry experiences a margin squeeze as we saw with the PC industry. In such an environment process managed enterprises can start to collaborate and outsource some of their secondary processes. Collaboration can take several forms like joint marketing, sharing back-office functions, etc with partners and suppliers. Process outsourcing for cost effectiveness is hot today as we see companies outsourcing call centers and customer support processes to other vendors. This is possible if companies are process managed and can change fast.

Flextronics

Flextronics focuses on delivering operational services to technology companies. Flextronics is an evolving chain and provides end-to-end operational services- including innovative product design, test solutions, manufacturing, IT expertise, and logistics to customers which include Microsoft, Xerox, Nortel, Motorola etc. Not many people know but Flextronics produces thousands of high-tech products including cell phones, telephones, computers, routers, and medical instruments—and provides a large portion of the manufacturing muscle behind leading companies such as Microsoft, Ericsson, and Xerox.

Flextronics is at the forefront of a being a process managed enterprise and its processes easily embed into Customers processes. The company gives its customers the ability to focus on their core competencies in such areas as research and development, design, and marketing with the confidence that their end products will meet the highest standards of quality. Flextronics has designed its processes to give its customers real-time control, monitoring and collaboration in services. This process managed machine is on a relentless drive to offer services to it's customers in real-time, with high quality and consistency and with great speed.

Here is a copy of the press release by Microsoft (http://www.microsoft.com/presspass/press/2005/feb05/02-14FlextronicsPR.asp). The press release clearly shows how Flextronics is using its innovative design process, manufacturing and assembly process to provide an end-to-end solution for customers. Author Jeff Ferry writes "while *many companies have downsized, spun off divisions, or otherwise shrunk, Flextronics has gone in the opposite direction — vertically integrating, globally expanding, and growing by acquisition"*. It is pretty clear to survive and thrive in an ever changing market enterprises need to manage their end-to-end business processes.

> **Flextronics and Microsoft Announce Windows Mobile-Based ODM Phone Platform for OEMs**
>
> CANNES, France -- Feb. 14, 2005 -- At the 16th annual 3GSM World Congress today, Flextronics (NASDAQ: FLEX) and Microsoft Corp. (Nasdaq "MSFT") announced Peabody, a GSM/GPRS mobile phone platform designed and integrated by Flextronics based on Microsoft® Windows Mobile (TM) software. Available only to Original Equipment Manufacturers (OEMs), Peabody is a low-cost, feature-rich Original Design & Manufacture (ODM) platform developed by both companies. Peabody can be brought to market quickly, cost-effectively and at high volumes.
>
> Flextronics is one of the largest mobile phone manufacturers worldwide and is the leading Electronics Manufacturing Services (EMS) provider. The company's extensive ODM mobile phone development experience makes Flextronics a strategic partner for the creation of a complete ODM mobile phone platform, such as Peabody.
>
> "As mobile phones become increasingly customized consumer products, OEMs are under pressure to produce low-cost, feature-rich phones that address the demands of this growing market," said Tom Deitrich, vice president of ODM Products, Flextronics. "Through ODM mobile phone platforms such as Peabody, Flextronics is providing OEMs with a cost-effective way to quickly expand their product lines and respond to market demands." He added: "Flextronics designed the Peabody platform based on Windows Mobile software, which delivers the substantial functionality required by OEMs while providing the flexibility for quick customization."

Press Release by Microsoft

Eugene McCabe, senior vice president of worldwide operations at Sun Microsystems, explains the attraction of Flextronics' integrated model for tech companies. After moving aggressively in recent years to outsource 90 percent of its production, Sun is today a "virtual manufacturer," taking orders from customers, configuring systems, and then handing the orders off to external manufacturing partners for production and shipping. "The only part of the order we touch directly is the information," says Mr. McCabe. It is pretty evident that Flextronics being a process managed company is thriving in an environment of change. It is enabling the next generation virtual enterprises.

UPS (United Parcel Service)

Some organizations on the cutting edge of process management are creating new industries out of internal functional units. UPS has created a new industry in Logistics Management. Intense specialization is a by product of being process managed.

Process Managed companies like UPS
are creating new Industries out of internal processes.

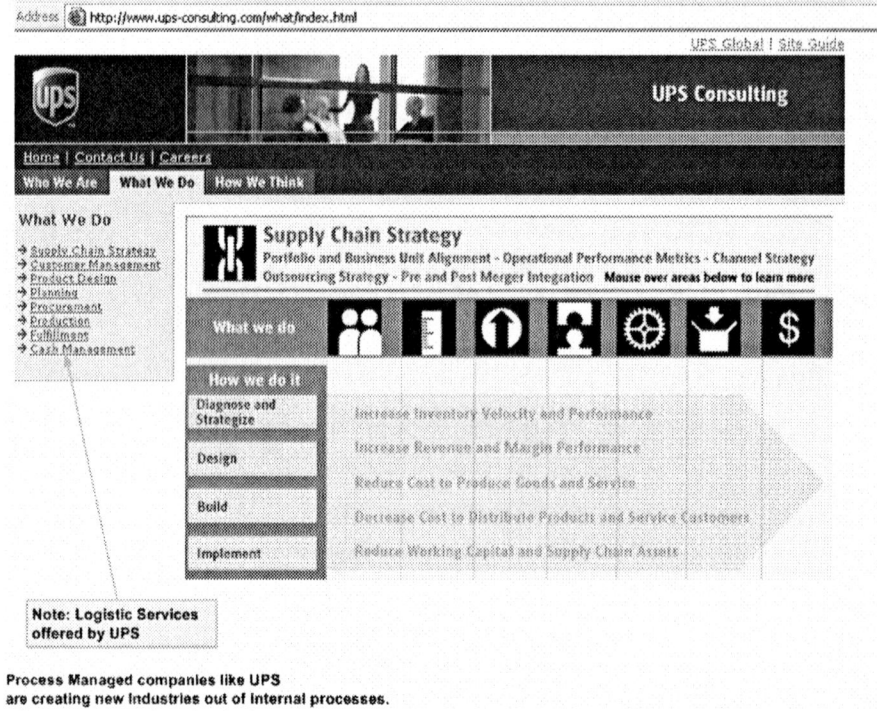

Fig 11.2 UPS is a logistics company

UPS has created a new industry in Logistics Management. Logistics used to be an internal business unit at UPS, but UPS is converting it into a new solution industry. UPS has transformed from a package delivery to a logistic management company. That happened because UPS understood that for their customers real-time information was becoming as important as their packages. In Figure 11.2 which is a snapshot of www.ups-consulting.com shows the various solutions and services which UPS offers. When you browse through the different services you will notice UPS has capitalized on its internal business processes and turned them into a new industry.

> Process managed enterprises are in a better position to capitalize on their core competency and extend their product and services portfolio.

Toshiba realized that laptops that were being returned for repairs, Toshiba was spending most of the time in logistics i.e. shipping laptops to a Toshiba repair center, getting the broken parts shipped to the repair center from various suppliers and shipping the laptop back to the customer.

As a result the turn around time was in weeks. So Toshiba outsourced the logistics and laptop repairing process to UPS. Certified technicians employed by UPS perform the repairs at a UPS center and customers now get their laptops back the next business day. In the PC World article titled "Toshiba Notebooks Get a Fix From UPS" Edward N. Albro writes *"Supply Chain Solutions, the business unit working with Toshiba, has been handling electronics equipment repairs for at least four years, according to Orzy Theus, spokesperson. UPS technicians repair some cell phones, printers, scanners, projectors, and personal digital assistants under similar arrangements with several other manufacturers, Theus says"*. Toshiba benefits in the following ways from UPS expertise

- UPS logistical capabilities mean shorter waits for parts to arrive and greater turnaround time for customers
- Toshiba can use UPS's sophisticated system for collecting data about transactions to track what parts are causing the most problems with the company's laptops.

Appendix A

BPM product greatest needs (to build Enterprise strength BPM systems)

BPM products are still evolving. I have built enterprise strength mission critical Business Process Management Systems (BPMS). From my 3.5 years of design, implementation and process management (design, deployment, monitoring, exception management and optimization) experience I have compiled a list of must have BPM product features and capabilities.

- **Table 1**: Is a high level list of capabilities, services and features a enterprise BPM product must have. These features are absolutely needed to create and support enterprise strength business processes. The list is a compilation of the features I have mentioned in the various chapters in the book.
- **Table 2**: Is a list of features a graphical process designer must have. Detail explanation, best practices and use each feature is mentioned in Chapter 4 (Business Process Design).
- **Table 3**: Is a list of features a BPM product must offer from a process deployment perspective. Detail explanation, deployment strategy, best practices and techniques of each feature is mentioned in Chapter 5 (Business Process Deployment).
- **Table 4**: Is a list of features the BPM product and process monitoring and exception handling tools must collectively provide from a live process management perspective. Detail explanation of business process monitoring and exception management is mentioned in Chapter 6 (Business Process Monitoring and Exception Management).
- **Table 5**: Is a list of features the process analytical reporting and performance metrics tools supplied with the BPM product must have. Detail explanation is in Chapter 7 (Process Optimization).

BPM Product		
Capability/ Feature	Description	BPM Product Offering
Graphical Process Designer	A single tool that can be used by different audience- business and IT for designing the business process.	Built in graphical process designer is a must have.
Business Rules Engine	Ability to separate business rules from business process.	1) Built in BRE (Business Rules Engine). 2) Ability to integrate with

		Business rules can be maintained by business in real-time.	third party BRE is a nice to have.
Deployment Tools		Deploy new and updated business processes to the BPM server.	Built in Process Deployment and management tools.
Monitoring and Exception Management.		Business process monitoring for process visibility and exception management.	Built in Process monitoring and exception management tools.
Process Analytics and Performance metrics		Process optimization is a continuous process in BPM.	1) Built in process report and dashboard view creation tools. 2) Publish process data structure so that process data can be merged into enterprise warehouse for complex OLAP style reports if needed.
User Directories		Work is routed to humans based on user directory information.	1) Built in organization chart management within BPM product is a must have. 2) Also support existing enterprise wide user directories like LDAP, Active Directory etc
Portal		Human participants manage their work items from a portal.	1) Built in thin portal should be core component of BPM product 2) Also integrate with leading industry portals, so that companies can leverage existing investment
Document Management		Part of being a process managed enterprise is the digitization of process supporting documents.	1) Provides built-in support for secure and reliable document routing. 2) Also integrate with existing DMS (Document management systems). so that companies can leverage existing investment
Enterprise Application Integration		Existing applications are participants in process orchestration.	1) Built in integration capabilities, services and web services support. 2) Seamless integration with

Appendix 138

		existing EAI products.
Collaboration	Built in collaboration technology for human collaboration.	Integrate with leading industry standard collaboration software.
BPM Server	This is the kernel of BPMS. BPM server orchestrates the business process. (Chapter 10)	Nice to have BPEL support.
BPM Administrator	Product specific administrator interface.	

Table 1 : Features needed to build enterprise strength BPM systems. Features needed from a process life cycle angle.

Graphical Process Designer (Chapter 4 in book)	
Capability/ Feature	**Description**
Intuitive process flow design	• Easy and Intuitive process design.
EAI activities	• Built in integration capabilities • Web services as activities within the process design.
Multiple Form Support	• Built in form designer. • Support for industry standard forms like ASP, JSP, infopath, DHTML etc.
Nested Model Support	• Capability to embed and invoke other business processes.
Collaborative process design	• Multiple BPM team members can work on one business process design concurrently. • Activity re-use by storing activities as objects.
Process Simulation	• Running the process design in a controlled environment for modeling and validation.
Event Notification	• Relaying events to appropriate audience in a declarative manner.
Process Documentation	• Generate process documentation out of process design.
Multiple design repositories.	• Ability to store process designs in separate repositories.

Table 2: Features needed to design complex business processes.

Appendix 139

Process Deployment Tools (Chapter 5 in book)	
Capability/ Feature	Description
Process Deployment	• Easy and secure process deployment with click and publish interface.
Side-by-side Version support	• Ability to orchestrate multiple versions of the same process. The latest process version will be used for all new process initiation.
In-Flight Process design update	• Ability to update running instances of business process.
Process Availability	• Ability to make business processes unavailable for initiation for a specific period of time.
Multiple deployment repositories.	• Ability to deploy business processes to separate repositories.
Process Initiation	• Ability to initiate business process using multiple ways – periodic, via email, via xml, via telephone, etc.

Table 3: Features needed to deploy complex business processes.

Process Monitoring and Exception Handling Tools (Chapter 6 in book)	
Capability/ Feature	Description
Process Instance (live or historical) View	• Ability to view process instance graphically in real-time.
Process Data	• Ability to view process and activity data as participants saw it.
Reports and Dashboards	• Ability to quickly build ad-hoc reports and dashboard views representing different process metrics.
Process Replay	• Ability to graphically replay a process instance.
In-Flight Process data update	• Ability to update live process data and process execution paths.
Workload Management	• Runtime task assignment

Table 4: Features needed to monitor business process and manage exceptions.

Appendix 140

Process Analytics and performance metrics Tools (Chapter 7 in book)	
Capability/ Feature	Description
Process Metrics	<ul><li>Display process metrics in a graphical fashion using dashboards and process reports.</li><li>Enable fast, easy and low-cost creation of process performance reports and dashboard views.</li></ul>

Table 5: Features needed to optimize business processes.

References

1) http://www.itweb.co.za/office/flextronics/pressclipping2.htm
2) http://www.strategy-business.com/media/file/sb37_04408.pdf (Flextronics Staying Real in a Virtual world) by Jeff Ferry
3) www.flextronics.com
4) www.ups.com
5) http://www.itweb.co.za/office/flextronics/0502150834.htm
6) http://www.itweb.co.za/office/flextronics/pressclipping2.htm by Jeff Ferry
7) http://www.microsoft.com/presspass/press/2005/feb05/02-14FlextronicsPR.asp
8) http://www.pcworld.com/news/article/0,aid,115907,00.asp - Toshiba Notebooks Get a Fix From UPS
9) http://www.forbes.com/2005/04/01/cx_pp_0401techsupport_print.html "Slashing Customer Service Costs" – Penelope Patsuris
10) http://www.forbes.com/cionetwork/2005/04/01/cx_pp_0401techsupport.html?partner=yahoo&referrer=
11) http://forums.us.dell/supportforums?ck=mn
12) http://developer.ebay.com/devprogram/whitepaper_eBayWebServices2.pdf
13) http://developer.ebay.com/devprogram/eBayBusiness.pdf
14) http://www.dmreview.com/editorial/dmreview/print_action.cfm?articleId=1011028
15) J. W. Wesner et al, Winning with Quality, Addison-Wesley, Reading, MA, 1995
16) Business Process Management – A Practicle Guide – Rashid N Khan
17) http://www.balancedscorecard.org/ The balanced scorecard institute
18) Gartner Group   http://www4.gartner.com/research/special_reports/Fusion.jsp Gartner October 2003 , Note Number : AV-20-9895
19) Business Process Management – The Third wave page 233 and 234 .
20) http://www.businessrulesgroup.org/brgdefn.htm
21) The third wave – business process management http://www.fairdene.com/processes/April2002-BPM-4thTier.pdf
22) Paper in Harvard Business Review (May 2003) title "IT Doesn't Matter" by Nicholas G. Carr
23) http://www.cioinsight.com/article2/0,1397,1455103,00.asp
24) Intangible Assets: Computers and Organizational Capital by Erik Brynjolfsson, Lorin M. Hitt, Shinkyu Yang
http://ebusiness.mit.edu/research/papers/138_Erik_Intangible_Assets.pdf
26) IT Doesn't Matter – (Harvard Business Review article by Nicholas G. Carr
27) *Business Process Management (BPM) is a Team Sport: Play it to Win! by Andrew   Spanyi*
28) The Agenda – Michael Hammer
29) http://www.cio.com/archive/071503/levis.html

Index

A
actionable Business Intelligence......29
actionable Data........................104
activity Based Costing 102
alerts...................................62
amazon.................................15
analytical 87
asynchronous...........................53
automated Activity.....................56

B
Balanced Scorecard 35,97
batch extraction 88
batch jobs............................108
business analyst........................49
business values...........19,20,25,39,81
BPEL.................................124
BPM............................19,20,31
BPM infrastructure....................41
BPM team......................40,47,48
BPM terminologies................39,51
BPM Pilot......................68,69,78
BPM environment....................70
BPMN................................52
BRE..............................43, 128

C
Collaboration...44,45115,116,117
Competition...........................15
Cost effectiveness metric.............102
Customer...............................13
Cluttered business process............66

D
Dell.......................15,16,18
Deployment 66
Document.............44,116
DMS................... 44,116
Discussion threads.....115,116

E
Ebay 70, 112
Efficiency..............13,14, 15
Event Notification 45,62
Exception Management.....88
Extended Enterprise.......15
EAI............................105

F
Flextronics 132
Fingar, Peter...............15, 45

I
Innovative........112,115,132
Instant messaging......38, 120
Interdependent Activities....66
In-flight process upgrade.....75
In-flight process change 65,66

K
KPIs 98,99

L
Loose coupling.........50

M
Management Services....128
Mindset....................19,22
Model......................14,15

N
Negotiation.................66

O
Opaque 29
Operation..................75, 77
Optimization..........41,85,94
Orchestration services   ...124

Index

P

point-to-point	25,107
portal	41
process activities	52
process chain	89,90
process discovery	49
process design	49
process documentation	63
Process downtime	75,76
Process efficiency	16,100
process instance	51
process managed	19
process metric	85,86,87
Process monitoring	39,41,81
process oriented	22
process participants	30,52
process repository	40
Process visibility	67,91,101

R

real-time	15,120
recipient	73,89,118
resource planning	25,103
return work	121
rigid	25,26
routing	58,59

S

sarbanes-Oxley	12
security	56,79
self service	13
Side-by-side process	64,65,75
Six Sigma	19,35
Smith, Howard	25,32,45
SOA	110
SOAP	39
split	64
stalled	52
state management	125
Sub process	64
synchronous	53,60

T

Transaction services	126

U

UPS	134
User directories	42,57

V

Visibility	20,45
Versioning	60,64

W

Wal-mart	14,16,18
web services	64
workload management	87
workflow	36
work in parallel	64
work item	53
work list	53

About the Author

Sandeep Arora has architected, designed and implemented enterprise strength mission critical business process manage systems, EAI projects, large scale N-tier web based projects, SOA based systems and client server applications. Sandeep has worked for Fortune 500 companies like Lockheed Martin, AMS, Pitney Bowes, Swiss Bank (now UBS) and Avon Products. Sandeep currently works for FM Global.

Sandeep got his Engineering Degree from the prestigious IIT (Indian Institute of Technology) Kharagpur, India.

Printed in the United Kingdom
by Lightning Source UK Ltd.
108448UKS00002BA/88